The Great Northern Kingdom

LIFE IN THE BOREAL FOREST

Text and Photography
by Wayne Lynch

Assisted by Aubrey Lang

**Fitzhenry &
Whiteside**

For my wife, Aubrey,
the best of the better
half of humanity

Fitzhenry & Whiteside acknowledges with thanks the
Canada Council for the Arts, the Government of Canada
through its Book Publishing Industry Development
Program, and the Ontario Arts Council for their support
in our publishing program.

THE CANADA COUNCIL | LE CONSEIL DES ARTS
FOR THE ARTS | DU CANADA
SINCE 1957 | DEPUIS 1957

Printed in Canada
01 02 03 04 05/5 4 3 2 1

National Library of Canada
Cataloguing in Publication Data

Lynch, Wayne.
    The great northern kingdom

    Includes index.
    ISBN 1-55041-617-0

    1. Taiga animals. 2. Taiga ecology. 3. Taigas.
    I. Lang, Aubrey. II. Title.
    QL112.L96 2001   591.73'7   C2001-910449-9

U.S. Cataloging-in-Publication Data
    (Library of Congress Standards)

Lynch, Wayne, 1948–
    The great northern kingdom : life in the boreal forest /
    text and photography by Wayne Lynch ; assisted by
    Aubrey Lang. — 1st ed.
    [160] p. : col. photos. : map ; cm.
    Includes bibliographic references and index.
    ISBN: 1-55041-617-0
    1 Forest ecology. 2. Taiga ecology. I. Lang, Aubrey,
    1947– . II. Title.
    577.37   21   2001   CIP

Fitzhenry & Whiteside Limited
195 Allstate Parkway
Markham, Ontario
L3R 4T8

In the United States:
Fitzhenry & Whiteside
121 Harvard Avenue
Suite 2
Allston, Massachusetts
02134

www.fitzhenry.ca

Contents

Acknowledgments

A book examining the largest terrestrial ecosystem on the planet should not be easy to write, and this one certainly was not. I began photography for the book in March 1984. During the first eight years of the project, I was the photographer for the book, and someone else was supposed to write the text. However, after many setbacks, the project collapsed in the early 1990s, and I thought it was finished. Two years passed, and I returned to the book with a new approach and vision. This time I would write the text myself and shoot an entirely new collection of photographs. I'm very happy I did, for this is probably the most important book of my twenty-year career. It is certainly the one I will remember the most.

Such a lengthy project could only have been completed with the generous help of many. I would like to thank, in alphabetical order: Dr. Rick Bonar, a pileated woodpecker researcher; Dr. Frank Broomhead, a skilled pilot and former medical colleague who flew me over countless frozen lakes, searching for wolves; Dr. Gordon Court, a good friend and helpful fellow photographer, with whom I've spent many enjoyable days in the field; master owl bander Ray Cromie; toad wrangler and outdoor buddy Hälle Flygare; Joyce Gould, the best doggone bog slogger in Alberta; Deidre and Dr. Graham Griffiths, who took me tiptoeing through the orchids in a secret fen; ecologist Dr. J. David Henry, whose vast knowledge of the taiga helped me to better understand this complex ecosystem; filmmaker Albert Karvonen, who hired me as a scientific advisor and scriptwriter for his documentary film *The Great Northern Forest*; Alberta forester William Mawdsley; the Mistik Management Limited of Saskatchewan; the friendly and capable folks at Calgary's Nova Photo; the cooperative staff of Saskatchewan's Prince Albert National Park, including superintendent Bill Fisher and wardens Dan Fransden and Lloyd O'Brodovich; researcher Dave Stepnisky, who gave me the lowdown on birds, beetles, and burns; Rick Strictland at the Alberta Provincial Fire Center; woodpecker researcher Jody Watson; bird man Tom Webb; and Alberta beetle specialist Mike Undershultz.

The Canada Council for the Arts provided generous funds to help pay the bills while I buried myself in the library. I am also grateful to forest fire specialist Brian Stocks for discussing his hot topic with me, sending me technical papers, and kindly providing the photograph that appears on page 24.

This is my third book with publisher Fraser Seely, managing editor Charlene Dobmeier, editor Liesbeth Leatherbarrow, and designer John Luckhurst. As before, I am thankful for their laughter, candor, and creative involvement.

Once the book was written, I surrenderd the text to five brave souls for technical review: wildlife biologist Dr. Gordon Court, botanist Joyce Gould, wildlife biologist Garry Hornbeck, naturalist Bradley Muir, and ecologist Dr. Fiona Schmiegelow. All were gentle in their criticisms and constructive in their suggestions, and I thank them for their attention to detail. Of course, I alone accept all responsibility for any errors that may have crept into the text. Dr. Schmiegelow also agreed to write an afterword, reviewing the tough issues facing the taiga of tomorrow. I'm grateful that she did.

Celebrated musician Carlos Santana also deserves special thanks. I always listen to guitar music when I am writing a new book, and because I usually play just two or three CDs over and over again, they soon become strongly associated with the literary process. Santana's latest album, *Supernatural*, was a real inspiration.

Finally, I wish to thank my loving wife, Aubrey Lang, to whom this book, and every other book I've ever written, is dedicated. Last summer I spent six weeks alone with her, camping in the Canadian high arctic. After twenty-five years of marriage, she is still the most stimulating, interesting, warm, and kind-spirited person I know. Life with her has been an exciting adventure.

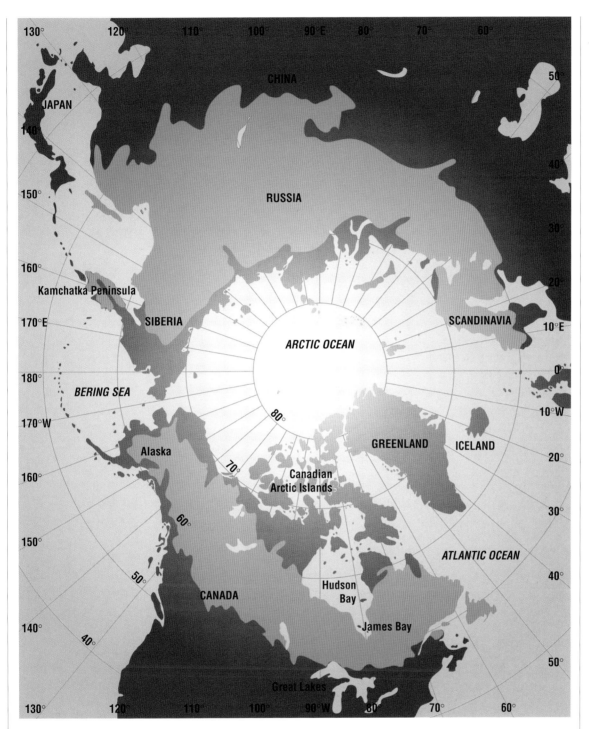

The boreal forest is the largest terrestrial ecosystem on Earth.

The taiga's emerald crown of conifers measures 12 million square kilometers (4.6 million square miles), roughly one and a quarter times the size of either the United States or Canada.

*A male ruffed grouse may
use the same two or three
drumming sites for as long
as it lives.*

Introduction

ON MY BIRTHDAY IN APRIL 1997, I MADE THIS ENTRY IN MY JOURNAL. This morning, I was inside the blind by 5:30 AM. The male ruffed grouse has become very tame. Earlier, as I tramped around outside the blind tightening the ropes, the defiant bird stayed on his drumming log a mere 10 meters away. Moments after I arrived, the spring sunrise fingered through the aspens and golden rays dappled off the chest of the drummer. The bird is now drumming every five to eight minutes, and with each session, the dried leaves at the base of the log flutter lightly in the wind generated by the blur of his wings. It's a glorious morning to be immersed in nature. The great northern forest around me is full with the rich songs of spring, and each song tells a tale—the rattle of an irate kingfisher as it chases an intruder from its lakeshore territory, the bugling music of a skein of migrant sandhill cranes as they wing overhead, the bold tattoo of a pileated woodpecker as it hammers on the crown of a dead tamarack in the nearby bog, the chatter of love-hungry wood frogs, and the whistling wings of a pair of common golden-eyes as the male doggedly shadows every dodge and swerve of his fast-flying mate.

Above: A cluster of black spruce clings to a rocky outcrop in the East Arm of Great Slave Lake, NWT.

Right: A wolf spider and a hitchhiking scarlet mite sun on a forget-me-not.

The body heat of an observer is enough to shake a spider web and dislodge its delicate decoration of dewdrops.

I never tire of spending time in the boreal forest. It's always memorable, always rewarding. It's the best place to celebrate a birthday. And as a wise friend once said to me, "It's a place to discover the hidden beauty in common things."

I have been enchanted by the boreal forest for as long as I can remember. During my childhood, my family often visited relatives in northern Ontario. It was there that I discovered the rich, intoxicating smell of a spruce forest after a summer shower and saw my first black bear as it deftly plucked juicy blueberries from the bushes beside a thundering waterfall. Later, as a university student, the forest lured me to the winding waterways of Quebec's La Vérendrye Wildlife Reserve for my first overnight canoe trip. Three days of slippery portages, rough water, and aching shoulders left me with indelible memories—the delicious sparkle and gurgle of gin-clear waters as they lapped against the granite boulders of an isolated lakeshore, the breathless whisper of the wind in the crowns of a cluster of

jack pines, and the haunting wail of a solitary loon as it mourned the dawn through a golden veil of swirling mist.

Several years later, in my first year of medical practice, the boreal forest summoned me again, and I moved to the small logging town of Chapleau, in northern Ontario. While there, I escaped into the surrounding forest as often as I could to soothe my psyche and to steal treasured glimpses of wolf packs crossing frozen lakes in search of winter-weary moose. In the summers, I trained my ear to the ethereal music of thrushes, vireos, and wood warblers, as they advertised their presence with the boldness of eagles, and savored the attention of inquisitive gray jays when these incurable thieves loitered to purloin a tidbit of my lunch. Eventually, in 1984, eight years after I left Chapleau and five years after I left behind my career as an emergency physician in Ottawa, I returned to the boreal forest. This time, however, I came to celebrate my infatuation and fascination with the place and to begin working on a book in its honor. *The Great Northern Kingdom: Life in the Boreal Forest* is my tribute to this seductive swath of spruce, pine, fir, larch, aspen, and birch—arguably the largest terrestrial ecosystem on the planet.

The boreal forest is an immense tract of trees that completely encircles the northern pole of the planet. Canadian ecologist Dr. J. David Henry aptly describes the forest as "Earth's evergreen crown." It's no surprise that within this vast forest there are countless stories to be told, and doing them justice would require a text many times longer than I wish to write or that you would find interesting to read. As a compromise, I have chosen my personal favorites—the best of the boreal, so to speak. Consequently, *The Great Northern Kingdom* focuses primarily on the living, breathing creatures of the forest. The book consists of twenty-seven essays that explore the unique strategies wild animals have evolved to survive in the boreal environment, and where possible, highlighting exciting new discoveries in animal behavior, ecology, and physiology. My wish is to tease you with the "gee whiz" science of the boreal forest, not bore you with its infinite minutiae. So, if you want boreal geology and soil analysis, or lengthy discussions on northern meteorology and water cycles, you should read a different book. But if you want to discover why muskrats purposefully leave trails of bubbles under winter ice, or how maggots, mites, and mosquitoes scavenge together inside carnivorous pitcher plants, or how daylight and diet control the pregnancy of a mother black bear, then read on.

The boreal forest is a circumpolar environment shared between North America and Eurasia. Although the tree species vary, forests on both continents are dominated by conifers, those needle-bearing trees with cones, and look quite similar. As well, many wildlife species are common to both areas. Even though I have traveled extensively from Alaska to Newfoundland, and from the southern fringe of the forest at the edge of the prairies to the northern tree line bordering the Arctic, my experience of the boreal forest is entirely North American. The text reflects this. Nonetheless, wherever possible, I have tried to infuse my discussions with references to northern Eurasia, to emphasize the common threads that link these two great, forested regions into a single, vast ecosystem.

The spirit and secrets of the boreal forest are not easy to uncover and understand; they are also not easy to capture on film. The northern forest is a wild and inspiring wilderness, but it can also be a dangerous, unforgiving place. In the past twenty years of my love affair with this forested enchantress, I was sometimes reminded of the fragility of life. Once in late September, my canoe overturned in a swirl of rapids, which transformed a casual outing into a frightening adventure. Another time, alone and on snowshoes, I broke through the thin ice of a beaver pond and risked lethal hypothermia. And yet another time, I became dangerously lost in a tamarack bog and learned how quickly terror can kidnap the mind.

These flirtations with tragedy only heightened my appreciation for the marvellous wonders and secrets the forest has shared with me—the plaintive howl of timber wolves, the thundering crash of aspens felled by burly beavers, the elaborate spring dances of fine-feathered grouse, and the heavy breathing of fire-eyed bull moose in full autumn battle. These are memories to cherish and relive, and these are memories to share. So, get comfortable, disconnect the telephone and computer, send the kids to the movies, and for a few hours let me introduce you to the subtle beauty and the rich biological complexity of the greatest forest of them all, the boreal kingdom.

Texture of the Taiga

The word boreal is derived from *Boreas*, the Greek name for the god of the north wind, so the boreal forest simply means the northern forest. Several other names are commonly used to describe this winter-weathered land: northwoods, great northern forest, and taiga (pronounced TIE-gah). Taiga is an interesting-sounding Russian word that originally meant "land of little sticks," referring to the stunted and fragmented character of the forest along its northern boundary with the Arctic tundra. Today the term taiga is used by scientists worldwide to describe the entire forest ecosystem, not just the tree-line portion.

In North America, the eastern boreal forest begins in the island province of Newfoundland, skips across to Labrador and northern Quebec, and then, in a band of spire-topped spruce and fir roughly 325 to 650 kilometers (202 to 404 mi.) wide, sweeps southward around the tip of James Bay (see map, page 5). The North American taiga is narrowest here, pinched between James Bay and the pink granite shoreline of Lake Superior. Here also, the forest reaches its most southern extent in the boundary waters of northern Minnesota. Heading west, the taiga widens again and cuts a broad course across the northern half of Manitoba, Saskatchewan, and Alberta, and then sweeps into Yukon. Just before this wide belt of conifers flows into central Alaska, a tongue of taiga splits off and follows the Mackenzie River north, beyond the Arctic Circle and almost to the river's outlet into the Arctic Ocean. This marks the northern limit of the boreal forest in North America.

Across the Bering Sea, which separates Alaska from Siberia, the boreal forest springs to life again about 560 kilometers (348 mi.) inland from the tip of the Chukchi Peninsula. The born-again forest quickly widens to consume eastern Siberia, as well as slivers of northern Mongolia and China. Here the taiga is a true giant of a forest, over 2,250 kilometers

The sugar content of many berries, including cranberries, increases once the berries freeze.

This wetland of golden sedges has been visited frequently by a rutting bull moose in northern Alberta.

The fleshy leaves of dog-tongue lichen are edible and tasty.

(1,400 mi.) wide. This is the cold, shadowed forest that the great Russian author Maxim Gorky called "the land of death and chains." The taiga of Siberia maintains its great width for nearly 1,930 kilometers (1,200 mi.) and then narrows as it crosses the Ural Mountains into northern Europe. The boreal forest finally ends in the fiords of western Scandinavia, halted by the salty spray of the north Atlantic. In its inexorable march across Russia, the taiga blankets nearly 80 percent of this, the largest country in the world.

Worldwide, the taiga's emerald crown of conifers measures 12 million square kilometers (4.6 million sq. mi.), roughly one and a quarter times the size of either the United States or Canada. The two land barons of the boreal forest are Russia and Canada, claiming 58 percent and 24 percent of this vast forest, respectively. Six other countries boast smaller tracts of taiga. The United States has 11 percent of the total, Norway, Sweden, and Finland just over 4 percent, and China and Mongolia, less than 3 percent.

Coniferous trees dominate the boreal forest. Across the breadth of boreal Eurasia about thirty types of conifers are found: eight species of firs (*Abies* spp.), ten species of spruce (*Picea* spp.), and a dozen species of stiff-needled pines (*Pinus* spp.) and feathery larches (*Larix* spp.). In comparison, a hectare (2.47 acres) of tropical rain forest may contain hundreds of different tree species, with the Amazon Basin, West Africa, and Borneo rain forests each boasting an entirely different and diverse confusion of trees.

In boreal North America, just five species of conifers comprise the bulk of the northern forests: black spruce (*Picea mariana*), white spruce (*Picea glauca*), tamarack (*Larix laricina*), jack pine (*Pinus banksiana*), and balsam fir (*Abies balsamea*). A single tract of North American taiga may contain all five types of conifers. Each tree species thrives under slightly different soil conditions, and together they partition the boreal landscape.

Sandy, well-drained soils are claimed by pines, which have a lengthy taproot to reach moisture deep underground. White spruce like upland soils that are well drained and where the permafrost, if present at all, is deeply hidden. Balsam fir prefers the same conditions as white spruce, but requires more rainfall. As a result, balsam fir thrives best in the moist eastern taiga of the continent. Tamarack is the only conifer in the boreal forest that is not evergreen; in autumn its soft needles turn gold and fall to the ground. This deciduous conifer, with its delicate wine-colored cones, grows especially well in cold, moist soils and bright sunlight. Consequently, tamaracks are often found in sphagnum bogs and swamps. The last of the taiga conifer clan, the widespread black spruce, is often relegated to lowland sites where the soil is thin, soggy, and poor in nutrients. When groves of these trees are underlain by permafrost, heaving ground causes the shallow-rooted spruces to tilt in different directions, creating the appearance of a "drunken forest."

Traditional peoples of the taiga, perhaps impressed by the tenacity of the black spruce, concocted many different medicines from the tree's branches, needles, and cones. The Woodland Cree and Chipewyan boiled black spruce cones and used the solution to relieve sore throats, mouth infections, and diarrhea. The Dene used the resin from saplings to treat snow blindness and painful corneal abrasions, and the Ojibwa dusted powdered needles onto cuts, blisters, and burns.

There are good reasons why conifers dominate the taiga. Noted scientist and writer Dr. E.

C. Pielou describes conifers as "waste not, want not trees" because they are so well adapted to conserve nutrients. Conifers retain their needles for three or four years, and at times for up to eight or nine years. Since new needles are not grown each year, fewer nutrients are required from the cold, impoverished, acidic soils in which these trees often grow. With needles ever-present, the tree can begin the crucial process of photosynthesis early each spring, and carry on late into the autumn, when warm weather permits.

Winter is the longest season in the boreal forest; for six months or more each year, the roots of taiga conifers are locked in frozen soil, unable to absorb water. Even so, the trees continue to lose moisture through the surface of their several million needles throughout the winter. If the water loss is too great, a conifer can die from desiccation. To combat this, conifer needles are covered with a protective, waxy coating that retards evaporation. Another moisture-saving strategy involves the pores on the needles' surface, through which the tree absorbs most of the carbon dioxide required for photosynthesis. The pores, more correctly called *stomata*, are sunken in pits instead of being flush with the surface, as they are on deciduous leaves. This adaptation reduces the desiccating effect of the wind.

A cursory understanding of the sex life of conifers is a good foundation for explaining some of the impressive winter movements of crossbills, grosbeaks, and chickadees within the boreal forest. All taiga conifers produce two distinct types of cones: male *pollen* cones and female *seed* cones. Both types occur on the same tree. The pollen cones are usually small and inconspicuous, and typically shrivel and fall from the tree during the summer. Female seed cones are what most of us are familiar with—the so-called pinecones of our childhood. Seed cones take a year or more to grow and mature. Usually they open in late summer and early autumn, and the tiny enclosed seeds fall out during winter or the following spring.

Not every year yields a good seed crop, which

is crucial for the seed-eating birds of the taiga. In some locations, white spruce and tamarack may produce a good seed crop every two or three years; in other locations or at other times, six years may elapse between such crops. The average black spruce yields a good seed crop just once every four years, and a similar timetable often applies to jack pines. Such variable cone crops mean cycles of boom and bust for feathered seedeaters.

During bust years, white-winged crossbills (*Loxia leucoptera*) wander nomadically across the taiga, searching for productive conifer cones. As many as ten thousand crossbills may move through an area in a single day. Biologists call these periodic migrations *irruptions,* and they occur at irregular intervals. In boreal Eurasia, white-winged crossbills irrupt every seven years on average and in the taiga of North America, every two to four years. Pine and evening grosbeaks (*Pinicola enucleator* and *Coccothraustes vespertinus*), as well as boreal and black-capped chickadees (*Poecile hudsonicus* and *P. atricapillus*) also experience periodic irruptions

feathery tufts of verdancy have delightful descriptive names such as small mouse-tail (*Myurella julacea*), fairy parasol (*Splachnum luteum*), electric eel (*Dicranum polysetum*), knight's plume (*Ptilium crista-castrensis*), rolled-leaf pigtail (*Hypnum revolutum*), curly heron's-bill (*Dicranum fuscescens*), pipecleaner (*Rhytidium rugosum*), and fragile screw moss (*Tortella fragilis*). Who says botanists are dull and unimaginative?

At the northern limit of the boreal forest, most mosses are replaced by lichens, those interesting amalgams comprising a fungus and an alga. The most conspicuous boreal lichens belong to the genus *Cladonia*, which consists of well over twenty different species. The common name for these lichens is reindeer moss. It was probably a know-nothing photographer who first applied the name, because the plants are not mosses at all, and every winter it's caribou, not reindeer, who consume great quantities of these ground-hugging plants in the North American taiga.

Although conifers are the dominant trees of the taiga, several kinds of deciduous hardwoods also thrive in the boreal forest. Dr. Pielou aptly calls these hardwoods "companions of conifers." In North America, the companion hardwoods include two groups of shrubs: willows (*Salix* spp.) and alders (*Alnus* spp.). They also include three main species of trees: aspen poplar (*Populus tremuloides*), balsam poplar (*P. balsamifera*), and paper birch (*Betula papyrifera*), the legendary tree of birch-bark canoe fame. All of these hardwoods are relatively fast-growing, sun-loving species. Alders and willows are the familiar bushy shrubs that choke the soggy shorelines of creeks and streams, the water-logged edges of bogs and fens, and roadside ditches. In contrast, birches and poplars prefer more elevated sites, where the soils are better drained and less soggy.

In early spring, all boreal hardwoods produce compact hanging clusters of miniature flowers, called *catkins*. As soon as the catkins ripen, they release a blizzard of tiny seeds. The seeds of birches and alders are equipped with broad

Moisture-laden air on a cold morning in early spring encrusts an aspen forest with hoarfrost.

when cone crops fail, but their numbers are generally not as impressive as they are for crossbills.

Within a coniferous forest, there is no leaf-free period when sunlight floods the forest floor and launches a showy burst of spring wildflowers, as happens in southern deciduous forests. Even so, as a nature photographer, one of the great joys of the taiga is to get down on my hands and knees to explore the lilliputian world of the forest floor. In the southern boreal, mosses are the primary ground plants. These

parchment wings, whereas those of willows and poplars have a wisp of cottony hairs to carry them aloft. It's no coincidence that the seeds of these plants are usually released before the leaves fully unfurl. In this way, the wind can freely disperse the seeds without foliage blocking the way. As a result, boreal hardwoods are common pioneer species in the taiga, ideally suited for colonizing newly exposed ground.

All boreal hardwoods modify local conditions to benefit the growth of the kingpin conifers. In winter, hardwood trunks frequently trap drifting snow. The deep snow may bury young conifers, buffering them from the worst of the cold and shielding their delicate branches from the abrasion of wind-driven ice crystals. In spring and summer, the leafy crowns of hardwoods create moist shady conditions on the forest floor, which are ideal for the growth of conifer seedlings. Then in the autumn, the fallen leaves decompose into nutritious mulch. In short, the hardwoods inadvertently serve as nurse trees for conifers.

The handsome aspen poplar is also known as the trembling or quaking aspen because of the way the wind causes its leaves to quiver with "fright." It is my favorite among the boreal hardwoods. To begin with, the two sexes grow on different trees—the arboreal version of sexual liberation. This means that a given aspen tree is either male or female, and its sex can be determined by the appearance of its catkins. This independent sexual lifestyle is not unique to the aspen. In fact, it is common to all poplars and willows.

I find aspens especially interesting because of the exciting discoveries researchers have made in the past fifteen years. Aspens can propagate by seed, as all trees do, but more often than not, they simply send out horizontal underground roots from which new trunks grow vertically. Thus, a unisex grove of a thousand aspen trees can actually be a single organism sharing a common root system and a common set of genes. Researcher Dr. Michael Grant at the University of Colorado believes a particular cluster of aspens in Utah is the largest plant, by weight, ever discovered. This aspen clone, as scientists call it, contains 47,000 tree trunks, covers 43 hectares (106 acres), and has an estimated weight in excess of 5.9 million kilograms (13 million lb.). That makes it three times heavier than the largest giant sequoia (*Sequoiadendron giganteum*), the former record holder. (For you trivia junkies, the heaviest organism with a beating heart is a 190,000-kilogram [418,000-lb.] female blue whale, *Balaenoptera musculus*.)

Aspen clones are not only big, they may also be old—very old. Individual aspens usually die within one hundred years, but the age of a single tree tells you nothing about the age of the clone. Researcher Grant admits he doesn't really know how long an aspen clone can live, but he guesses it may be longer than ten thousand years. Other less conservative scientists speculate an aspen clone can live a million years or more, making it "essentially immortal."

Once you know that aspen poplars grow in clones, you can confirm this fact for yourself with a stroll through the taiga. Because clone members have identical DNA, they are alike in a number of recognizable ways: similar angles between the trunk and the main branches, the same schedule of spring leafing out, and a similar timetable for the autumn flush of gold.

One final tidbit of aspen biology deserves a few words. Under normal growing conditions, every aspen tree in a clone continually flushes hormones into the common root system. This inhibits the growth of additional stems. When a tree dies, so does its hormone signal. Should a great many trees die all at once, as happens in a forest fire, for example, the resultant drop in hormones triggers a dramatic increase in the number of new trunks sprouting. In fact, the number of regenerating stems may far exceed those lost in the original fire. On a single hectare, researchers have counted as many as 988,000 new aspen stems (400,000/acre). Now that's "gee whiz" science at its best.

The Roving Forest

I remember sitting on the rocky shoreline of Hudson Bay one July day in 1995. A continual, icy wind blew off the water. This last breath of winter numbed my face and made my nose run like a sap spigot in spring. It didn't feel much like summer, but then again, it was 10:00 PM and I was on the northern frontier of the taiga. Offshore, the crimson orb of the sun hung low on the horizon, transforming the rippled waters of this vast inland sea into liquid streamers of crimson and gold. I felt quite insignificant, but that is how I should have felt. After all, at that moment I was an insignificant speck of humanity, barely half a century old, squarely planted on an ancient shoulder of Precambrian quartzite, six hundred million years old! I savor such moments of clarity and realization. They put my life in perspective.

Fifteen thousand years ago, at the height of the last glaciation, Hudson Bay was the center of the Laurentide Ice Sheet, which extended south beyond the Canadian-American border. At that time, most of the boreal forest was buried under tons of gouging, lifeless ice. Those portions of Alaska and Yukon that escaped glaciation were covered with tundra grasses and sedges. The taiga, or at least a version of it, only survived south of the continental ice sheet in the northern and central United States. In those days, the composition of the boreal forest may not have been the same as it is today, mainly because the climate was different. South of the frozen lip of the ice sheet, summers may have been longer and winters less protracted and frigid. Overall, it was likely a warmer taiga than the one we now enjoy, permitting trees such as ash (*Fraxinus* spp.) and elm (*Ulmus* spp.) to mix with the dominant conifers.

When the world began to warm up, the forest moved steadily northward. Under average

In North America, the taiga tree line meanders farthest north where it cuts across the delta of the Mackenzie River.

The polished granite shoulder of the Canadian Shield surfaces beside the Slave River in Wood Buffalo National Park.

conditions, plant life advances about 161 kilometers (100 mi.) every thousand years. The taiga's northward march may have been faster than this, since the scoured, denuded landscape abandoned by the glaciers would have presented little competition for pioneering plants. A branch of science called *palynology* uses ancient pollen grains dredged from the sediments beneath old bogs to determine which plants grew in a given area thousands of years ago. An experienced palynologist can determine the identity of pollen grains by their appearance under a microscope, but palynology is in no way an exact science. Experts may be able to tell the difference between the pollen of a pine, a spruce, and a birch. However, they cannot usually differentiate between white spruce and black spruce, nor between dwarf birch (*Betula pumila*), water birch (*B. occidentalis*), and paper birch. As a result, for these and other reasons, many of the descriptions of ancient environments are speculative.

The taiga's northern edge at the tree line reached its present position around 3,500 years ago and has remained there ever since. What stopped its movement? Intuitively, you might guess that extreme cold keeps the taiga from creeping farther north. Yet, black and white spruce, which are the guardians of the tree line, are quite resistant to cold and able to withstand temperatures below -62°C (-80°F). You might also guess that wind and permafrost are two other possible factors limiting the growth of trees. This time you would be right. The winter wind, burdened with crystals of ice, can prune and kill trees; what's more, the permafrost, which rises closer and closer to the surface the farther north you travel, forces trees to anchor themselves with roots so shallow that eventually they topple over. However, wind and permafrost are just minor players in determining the position of the tree line. The leading role belongs to summer warmth, for without a minimum summer growing period, trees simply cannot survive.

Much of the northern tree line of the taiga roughly parallels the 10°C (50°F) isotherm, a hypothetical line joining all points in the north where the average temperature of the warmest month of the year (usually July) is 10°C (50°F). This is the minimum July temperature required by the hardiest of trees to complete their annual growth cycle. If the July average is any colder, the trees will die.

The tree line is a blurry, fluctuating boundary. As global climates have changed, so has the position of the taiga's tree line. For example, during the so-called Hypsithermal Period, 8,500 to 5,500 years ago, the boreal forest was the warmest it has been anytime during the present interglacial period. At that time, the northern edge of the taiga was as much as 330 kilometers

(200 mi.) farther north than it is today. With the spectre of global warming looming on the horizon, the taiga may once again pick up its roots and move north.

The northern edge of the boreal forest merges with arctic tundra along its entire length, but the situation is more varied along the forest's southern boundary. In eastern North America, the southern edge of the taiga gradually merges with the pines, maples (*Acer* spp.), oaks (*Quercus* spp.), and beeches (*Fagus* spp.) of the Great Lakes–St. Lawrence forests. In the middle third of the continent, the taiga ends at the edge of the aspen parkland and fescue prairies. In the west, the forest quietly merges with the conifers of the mountains.

No discussion of the North American boreal forest is complete without mentioning the Canadian Shield. The shield is the geological core of North America and forms the basement bedrock under most of the continent east of the western mountains. This same block of bedrock also underlies many of Canada's arctic islands and virtually all of Greenland. Bedrock is dead rock, and indeed, the rocks of the shield have been dead for a very long time. Some may be as old as four billion years, making them among the oldest rocks on Earth.

Much of the southern and western portion of the Canadian Shield is buried beneath younger rocks. Only in the central and eastern areas of the boreal forest does this ancient bedrock lie hidden just beneath a shallow covering of soil or reveal itself at ground level as barren outcrop. The shield surfaces in the heart of the eastern taiga because this was the hub of glaciation, repeatedly scoured by the advance and retreat of gargantuan glaciers. Geologists would enthuse that the Canadian Shield is the largest area of exposed Precambrian rock in the world. As someone who is geologically challenged, I would rather describe the shield as pillows of polished rock that replay the echo of a yodeling loon, catch the firelight of a taiga sunset, and cradle the waters of innumerable tumbling rivers.

Ice and Fire

The taiga is a landscape forged from two elemental forces—ice and fire. There's a common expression that describes someone who barters one set of difficulties for another. We say they jump from the frying pan into the fire. The taiga, in its own way, jumps from the deep freeze of winter into the flames of fire, year after year.

Winter in the taiga is legendary. Worldwide, the boreal forest spans a great range of latitudes, from 45° to 70° North, and is classified throughout as cool and humid, with cold, snowy winters and brief, relatively cool summers. This terse description does little justice to the taiga's formidable winter disposition. For six to eight months each year in many regions of the boreal forest, snowfall greater than the height of a human buries and smothers the vegetation, warps and snaps the branches of trees, taunts and tests the wildlife. With the snow comes frigid cold—cold so painfully intense that sometimes the trees audibly wince in protest. The average January temperatures in boreal Manitoba range between -22°C and -30°C (between -8°F and -22°F). This is the boreal version of balmy when you compare it with the record low temperature for the taiga of Alaska, which was measured in Fairbanks at -54°C (-65°F). In Canada, the record low comes from Snag, Yukon, a small Native community, originally named for a nearby creek that was choked with dead trees. In February 1947, the temperature in Snag dipped (hell, it almost disappeared) to -63°C (-81°F). Perhaps, it was such a night when Robert Service scribed his famous poem "The Spell of the Yukon."

The winter! The brightness
  that blinds you,
The white land locked tight
  as a drum,
The cold fear that follows
  and finds you,
The silence that bludgeons
  you dumb.
The snows that are older
  than history,
The woods where weird
  shadows slant,
The stillness, the moonlight,
  the mystery,
I've bade 'em goodby—but I
  can't.

The boys in the bar in Whitehorse knew all about Snag and its climatological claim to fame. What they didn't know was that such temperature extremes are common in the taiga of eastern Siberia. In some winters, the *average* January temperature in Verkhoyansk is a vodka-freezing -56°C (-69°F). Russian author A. Borisov mentions in his book *The Climate of the USSR* the phenomenon known locally as "the whisper of the stars"—the soft rustling of ice crystals from exhaled breath as they fall on the snow.

In the taiga of North America, annual snowfall is least in the center of the continent and increases eastward and westward from there. Northern Saskatchewan, for example, receives up to 200 centimeters (79 in.) of the white fluffy stuff in a winter. Farther west, in Yukon, the forests receive at least as much snow as Saskatchewan does, but sometimes they are blanketed to a depth of 400 centimeters (157 in.). Going east, the boreal forests in Quebec receive up to 300 centimeters (118 in.) of winter whiteness, and Newfoundland may be buried with 400 centimeters (157 in.) of snow.

Most of us think of winter as just one of four seasons, but for some Native peoples of the taiga there are six seasons in a year. Native peoples have always lived close to the land, sensitive to the subtleties of its temperament and the rhythm of its changes. For the Woodland Cree of the taiga, summer does not ease into autumn, but flows instead into the season of *Mikiskaw*,

The graphic beauty of a lake in northern Manitoba is frozen after the first frigid night of autumn.

Fireweed, rich in vitamins A and C, was a traditional food of many Native peoples of the boreal forest.

or freeze-up. This occurs after the leaves have fallen and the lakes have been capped with ice. Between winter and spring, the Cree recognize another season, *Mithoskumin*, or breakup, when the land is finally freed from winter's frozen grip, and the rivers rush and roar again.

Ecologist Dr. David Henry wrote, "Mithoskumin is a time of real stress for the trees, particularly the conifers. Their needles and branches bask in 14 hours or more of daylight, and yet they gain very little moisture from the frozen soils. Furthermore, the lakes, the local rainmakers, are still sealed with ice." Henry describes these stressing weeks from early April to late May as a "season to burn" and one of the most flammable times in the taiga.

In any given year, tens of thousands of wildfires flare to life, blaze, and then cool to ashes

3,830 square kilometers (1,479 sq. mi.) of taiga forests. In Russia, during the same time period as reported for Canada, roughly 22,000 fires were recorded annually, burning approximately 11,198 square kilometers (4,324 sq. mi.) of forest every year.

Today, a third of all forest fires are caused by lightning and two-thirds by *Homo stupido*. However, lightning fires are responsible for 85 percent of the hectares burned, most likely because these fires are sparked in remote areas and are difficult to extinguish. As a result, lightning fires are, on average, ten times larger than those caused by human carelessness.

The 1998 fire season came rapidly on the heels of the worst El Nino in recorded history. In the preceding autumn and winter, this equatorial ocean phenomenon drastically altered global weather patterns, and the effects were felt as far away as the taiga. The resulting dry spring conditions proved how extremely combustible the taiga can be, when the circumstances are right. In the boreal forests of Alberta, there were 1,191 fires caused by lightning that year, a 250 percent increase over the average number during the previous nine years. Not only were there many more lightning fires, but they consumed a total of 6,516 square kilometers (2,516 sq. mi.) of forest, nearly fifteen times more than the average for those same years. One of the largest fires, the Virginia Lake blaze, burned 1,629 square kilometers (629 sq. mi.) of forest. Taiga fires are sometimes even larger than this.

The largest wildfire in Ontario occurred in 1948, near Chapleau. You will remember this was the town where I first practised medicine and which was one of my earliest boreal forest stomping grounds. The immense Chapleau-Missinaibi fire burned for over a month, charring 2,590 square kilometers (1,000 sq. mi.) of taiga. The blaze blackened the sky with thick clouds of smoke that drifted as far away as Texas, where streetlights were reportedly turned on during the day because of the smoke. One of the largest fires in taiga history occurred in 1995, in the Horn Plateau region of the Northwest Territories, northwest of Great Slave

across the length and breadth of the boreal forest. Brian Stocks, a senior research scientist with the Canadian Forest Service, provided me with some startling global statistics. In Canada, from 1981 to 1996, there was an average of 9,246 fires every year, charring roughly 24,992 square kilometers (9,650 sq. mi.) annually. In Alaska, between 1990 and 1996, there were roughly 670 fires every year, with an average yearly loss of

A raging crown fire consumes a forest of jack pines in northern Ontario.

Photo: Brian Stocks

Lake. The massive firestorm was ignited by a bolt of lightning and burned for four months, from June 2 to October 6, consuming 10,499 square kilometers (4,054 sq. mi.) of forest, an area nearly twice the size of either Prince Edward Island or the state of Delaware.

Wildfires are nothing new to the boreal for-est. The presence of buried ash in the thin taiga soils attests to the forest's longstanding familiar-ity with the incendiary powers of lightning. Ecologists estimate that the average taiga forest burns at least once every 150 years. The resinous branches and needles of conifers are highly flammable, and the arboreal lichens draping the branches of many of these trees make them all the more combustible. As well, the superficial roots of many conifers are easily killed and consumed in surface fires. Hardwood aspens (*Populus tremuloides*), willows (*Salix* spp.), and alders (*Alnus* spp.) are equally vulnerable to flames. Only the balsam poplar (*Populus balsamifera*) exhibits some resistance to fire. The corky bark on this tree can be 10 centimeters (4 in.) thick, shielding the tree from all but the most intense heat. Of all the taiga trees, the jack pine (*Pinus banksiana*) has the most fascinating adaptations to fire. Dr. David Henry jokes that the tough cones of the jack pine are "about as soft and flexible as a piece of iron ore ... and look like a hand grenade." In a jack pine, 10 percent of its cones open in late summer in the manner of most conifers. The remaining 90 percent are committed pyromaniacs, relying on the intense heat of a forest fire to melt the resin that seals their seeds inside. These pinecones can withstand temperatures of 700°C (1,300°F) for three minutes and still have all their seeds survive.

Once a fire sweeps through a forest, it leaves behind a death scene of skeletal trees and smoldering, mineral-rich ash. But all is not in mourning. The abundant sunlight and removal of competition makes the charred site a land grabber's free-for-all. After a fire in Ontario, grasses and sedges started to poke through the ashes within two weeks. Within two months, feathery fern fronds had unfurled (I dare you to say that three times quickly) and aspen seedlings were sprouting up as thick as the hair on a dog's back. In Alaska, foresters did a seedling count on a year-old burn. The final tally was 2,000 spruce (*Picea* spp.) seedlings per acre, 5,000 poplars (*Populus* spp.), and 800 birches (*Betula* spp.).

The succession of plant life on a taiga burn is highly variable. The bedrock composition, soil moisture, latitude, topography, and past vegetational history of the site combine with other factors to determine which pioneer plants will colonize the ashes and which will be the successors. Compare a simple spruce forest burn in Alaska with one in the shield country of northern Ontario. In Alaska, a burn site is initially dominated by fireweed (*Epilobium angustifolium*), followed by birch and aspen, and eventually spruce. In Ontario, fire-adapted jack pines are the early colonizers; they dominate a site for many years before spruce eventually regain their dominance.

In the late 1980s, I photographed a burn in northern Ontario, two days after it had been extinguished. Fallen tree trunks still glowed with embers, and funnels of smoke curled skyward from the warm, powdery ashes all around me. I knew how soon plants would invade this blackened forest, but I wondered how fast the critters would return. Affable graduate researcher Dave Stepnisky came to my rescue and gave me the lowdown on beetles, birds, and burns.

Wood-hungry beetles are generally the first insects to colonize a burn. Many possess infrared receptors on their legs that enable them to target the heat of a fire. Some species can detect a small 20-hectare (49-acre) burn, 4.8 kilometers (3 mi.) away, and a blazing fire at even greater distances. Some wood-boring sawyer beetles (*Monochamus* spp.) can even orient to the smoke plume issuing from a burn. Stepnisky explained that beetles start to arrive within days after a fire has passed and "some have been recorded attacking trees while they were still smoldering."

Woodpeckers and wood-boring beetles go together like biologists and beer. Burned coniferous forests are a magnet for two closely related species of northern woodpeckers, the black-backed woodpecker (*Picoides arcticus*) and the three-toed woodpecker (*P. tridactylus*). Blackbacks, the more specialized of the two, search most heavily on moderately to heavily burnt spruce, excavating larval, wood-boring beetles in the Family Cerambycidae. In contrast, three-toeds tend to forage on less-burnt spruces and search under the bark for the larvae of bark beetles in the Family Scolytidae.

As beetle numbers go, so go the woodpeckers. Beetles and woodpeckers use burnt forests heavily for up to eight years, but the numbers of both decline slowly in the following years.

*A four-month-old moose
calf follows its mother as
she crosses a shallow river
in Alberta.*

Right: This male pine grosbeak traveled in a winter flock of a dozen birds.

Winter

WHEN I THINK OF WINTER IN THE BOREAL FOREST, I think of the invigorating tingle in my cheeks and the purifying feel of cold clean air as it chills my face and fills my lungs. I think of snow and its dazzling whiteness, and how it glints in the moonlight and blankets the landscape with a peaceful purity. From winters past, I remember the fullness of a fox's tail as the animal flees across a whitened hillside, the fluid glide of a great gray owl as it hunts for voles hidden beneath the snow, and the frozen breath of a bison as it muzzles the ground for a meagre meal. I also recall how an aurora fades and brightens in a frigid sky, like the heaving of great lungs, how snowshoes swish in the silence of a winter sunrise, and how a flock of redpolls twitters with the resilience of life when painful cold threatens to congeal the forest.

For most of the animals and plants of the taiga, winter is the season of challenge—a time of testing, in which only the winners survive. It is the time when river otters belly through the snow, when insects hide in crystalline tunnels, and when chickadees chill their pounding hearts. These, and other stories, make winter a season of wonder.

Above: A female red fox is called a vixen, whereas a male is called a dog.

27

Life Under the Ice

For six to nine months each year, the taiga winter transforms the lives of the Canadian beaver (*Castor canadensis*) and the muskrat (*Ondatra zibethicus*) into a water world capped with ice, shadowed by deep snow, and easily depleted of food. Neither of these aquatic, roly-poly rodents hibernates for the winter, and each has its own strategies for surviving the challenging circumstances of the season.

As a tall naked ape, I was never designed to confront a boreal winter; however, I have always been intrigued by the ways that wild creatures cope with the cold, especially those that swim in freezing water. Both the beaver and the muskrat have luxuriant pelts with a thick dense underfur that traps an insulating layer of warm air next to their skin. But this is not news. For millennia, humans have coveted the furs of these animals. In North America, exploration of the boreal forest from the seventeenth to the nineteenth century was driven primarily by the European appetite for beaver hides, and the warmth and wealth they could bestow.

The muskrat and the beaver are master animal architects who build shelters to better their odds against the winter. Typically, the brawny beaver overwinters inside a conical lodge of branches, logs, and rocks, sealed and cemented with mud and vegetation dredged from the pond or lake waters that surround the lodge. A large beaver house can be more than 12 meters (39 ft.) in diameter at its base and reach 3 meters (10 ft.) in height. It always appears smaller than this because a large portion of the lodge is hidden underwater. Two or more submerged entrances lead to the single inside chamber where the beaver family crowds together. In Ontario, a lodge that housed three beavers maintained an inside temperature that hovered around the freezing point all winter long, even when the outside temperature dropped to -21°C (-6°F). Another beaver lodge in Alaska illustrates how great the thermal advantages of shelter can

sometimes be. There, the inside temperature of the snow-covered lodge rose as high as 6°C (43°F) on some winter days and cooled to -4°C (25°F) when the outside temperature plunged to a lethal -45°C (-49°F). Without the protection of a lodge, beavers would rapidly perish in such congealing temperatures.

A muskrat's lodge is much smaller than a beaver's, which is no surprise, since the average 20-kilogram (44-lb.) beaver is twenty times heavier than a muskrat. Unlike the beaver, which commonly occupies a lodge year-round, the muskrat builds a house just for the winter. Starting in September, the industrious little rodent stacks cattails (*Typha* spp.), bulrushes (*Scirpus* spp.), pondweeds (*Potamogeton* spp.), and mud into a conical mound up to 1.3 meters (4 ft.) high at the edge of its marsh domain. The thickness of the walls surrounding the single chamber inside is usually less than 0.3 meters (1 ft.), about one third the thickness of the walls of some beaver lodges. Even so, the frozen walls of a muskrat's winter house are vital protection against winter's severity. In a Manitoba muskrat study, the inside temperature of the houses averaged a life-saving 35 C° (63 F°) warmer than the outside air!

Both the beaver and the muskrat take advantage of warmth from fellow lodge-dwellers. Commonly, five or six beavers overwinter in the same lodge, but as many as twelve may snuggle together. Beavers that overwinter together are always related and usually include an adult pair, their latest kits, plus yearlings from the previous year. Muskrats also huddle to heat up. As many as half a dozen unrelated adults and juveniles may share the same winter house. For both muskrats and beavers, huddling conserves body heat and saves precious energy.

Both beavers and muskrats eat their way through winter, but they do it in different ways. A muskrat lives from day to day. It doesn't pad itself with a great layer of fat or store piles of

*A pair of muskrats
groom each other
on the ice next to
their winter lodge.*

surplus food. Instead, it feeds daily, foraging on the underwater rootstalks of horsetails (*Equisetum* spp.), reeds (*Phragmites* spp.), cattails, and bulrushes. Sometimes it may dig through 50 centimeters (20 in.) of mud to reach an edible root.

In contrast, beavers plan ahead. Beginning in late August, beaver family members build an underwater pile of freshly cut branches beside their lodge and continue to add to it, until ice freezes their world shut. In the southern taiga, beavers prefer aspens (*Populus tremuloides*) as

A fresh snowfall mantles the shoreline and distant lodge of a family of beavers in northern Saskatchewan.

A beaver gnaws on
a discarded aspen
branch during a
late-winter thaw.

their winter food, but farther north they tend to store willows (*Salix* spp.). Researchers in Wood Buffalo National Park, in the Northwest Territories, analyzed the winter food caches of three different beaver families. One family of five had a cache weighing 325 kilograms (717 lb.), half of which was alder (*Alnus* spp.), the other half of which was a mixture of willow and red-osier dogwood (*Cornus stolonifera*). Another pair of beavers had stored just 80 kilograms (176 lb.) of food, of which 87 percent was willow. Finally, a large family of eleven beavers had cut and cached 380 kilograms (838 lb.) of branches, nearly two-thirds of which was willow; the rest was a mixture of birch (*Betula* spp.), aspen, alder, and dogwood.

The beaver's winter food comprises a high-fiber, low-quality diet; it consists almost exclusively of the bark and small branches of the woody plants it caches. Beavers digest these meagre nutrients in three ways. First, the food is soaked in a strong acid broth in the beaver's stomach. Next, it is attacked by powerful microbes in the cecum, a pouch at the junction of the small and large intestine, which can digest up to 30 percent of the cellulose. Finally, the beaver eats its own feces, taking the soft droppings directly from its anus. In this way, the food passes through the animal's gut for a second round of digestion.

Most researchers conclude that no matter how much food a beaver family stores, or how well they digest it, the winter cache usually does not contain enough nutrition to sustain them for an entire taiga winter. The solution? Pack on the pounds in summer. Going into autumn, a well-fed beaver may be from 30 to 40 percent fat, half of which it stores in its leathery, paddle-shaped tail. Over the winter, the volume of its tail actually shrinks by half, as the beaver consumes the stored energy.

To reach their winter food, both the muskrat and the beaver must dive into frigid water under the ice and stay submerged for a length of time. On average, a beaver leaves the lodge once a day to forage from the family food cache and is gone from five to forty minutes. Along the Mackenzie River in the northern boreal forest, a beaver may leave to feed just once a week. Normally, the animal can hold its breath for up to ten minutes, so on longer trips, a beaver presumably scavenges naturally occurring oxygen bubbles trapped under the ice.

The story is different for the muskrat. Each autumn, this animal prepares for the winter challenges of life under the ice by increasing its diving limit, something the beaver does not do. To achieve this, the muskrat increases both its blood volume and the amount of oxygen-binding myoglobin in its muscles. These two improvements nearly double the muskrat's store of oxygen and enable it to stay underwater for up to 40 percent longer than it can in summer. Even so, the little rodent's diving limit is just fifty-eight seconds. In this time, a muskrat can swim underwater about 46 meters (151 ft.) before it runs out of air. There's just one problem. Winter feeding areas may be three or four times that distance apart. The muskrat solves this dilemma in two ways. In autumn, when the animal builds its winter lodge, it also builds a number of feeding shelters nearby. These shelters resemble a lodge, but are about half the size. After ice forms, muskrats build a third kind of shelter, called a *pushup*. The toothy rodent gnaws a hole through the ice and then stuffs soggy vegetation up under the snow where it freezes, forming a small cavity where the animal can rest and eat. These shelters are spaced closely enough so that the muskrat never exceeds its dive limit when traveling from one to another. Twenty-five years ago, veteran muskrat researcher Dr. Robert MacArthur observed, "Pushups and feeding shelters are often arranged about a central dwelling lodge in a stepping-stone pattern. Thus the residents have access to all points in their home range via a series of short underwater excursions."

In the 1990s, Dr. MacArthur described another tactic used by the adaptable muskrat to survive under the ice. During a typical dive, the animal may purposely exhale half the air in its lungs. Along frequently used routes, the exhaled air forms a trail of bubbles under the ice. MacArthur sampled the gas in these bubble trails and discovered that they contain between 8 and 20 percent oxygen. He also observed muskrats inhaling the bubbles through their nostrils. But why does the animal exhale the air in the first place? Why not hold onto it and eliminate the bother of snorting bubbles under the ice? The answer may be linked to carbon dioxide concentrations in the air.

Inside a crowded muskrat lodge, the carbon dioxide level may rise to 10 percent, a four-hundred-fold increase over its usual concentration in air. Thus, when a muskrat leaves the lodge, its last gulp of air may contain a high level of unwanted carbon dioxide. The solubility of carbon dioxide in water is thirty times greater than it is for oxygen. When the muskrat exhales underwater, the unwanted carbon dioxide rapidly diffuses from the gas bubbles and has disappeared by the time the rodent inhales the air again! Now that's delicious science.

Muskrats and beavers have one final strategy for foraging underwater: they turn up the heat. Normally, the moment either rodent plunges into the numbing 3°C (37°F) water under the ice, its body begins to cool. Neither animal lets its body temperature dip more than 2° below 37°C (3.6° below 98.6°F). Thus, their time in the freezing water is limited by how rapidly they cool. A muskrat has just fourteen minutes before it must find shelter and actively warm itself. The beaver, with a much larger body mass, cools three to four times slower than the muskrat, so it can stay in the frigid water longer. In both cases, however, the rodents may need more time to forage than the cold water permits. To prolong their time underwater, beavers and muskrats voluntarily raise their body temperatures a degree or two before they enter the water. The beaver's modest 0.6 C° (1 F°) rise in temperature gives the animal an additional fifteen minutes in the water. For the muskrat, a 1.2 C° (2 F°) boost in body temperature yields an extra nine minutes. Clearly, in a taiga winter, mere minutes sometimes determine whether an animal lives or dies.

Tactical Tits

It was February 11, in a cold spruce forest in northern Saskatchewan. A 10-centimeter (4-in.) snowfall burdened the trees and muffled their morning whisperings. The only sound was the rhythmic swish of my snowshoes. I had risen at sunrise to wander the deep snows of the taiga, unravel the stories written in tracks, and search for wildlife tenaciously challenging the day. I stopped on a ridge of jack pines (*Pinus banksiana*) to listen for life. Off in the distance, I detected the faint whistle of a gray jay (*Perisoreus canadensis*) and then there was penetrating silence. It's remarkable the way the brain works and immediately recalls a forgotten memory.

Often when I'm immersed in the gentle quiet of the wilderness, I think of a poignant conversation I had with a young German woman twenty years ago. She confided that before she had visited the taiga of Canada, she had never in the twenty-six years of her urban life known the peaceful sound of silence. After that, I never took such moments for granted. Within the hush of the jack pines that morning, I reflected again on the words of that woman, but the soft contact calls of a flock of chickadees gently interrupted my introspection. The tiny birds were never still, constantly peering into clusters of needles, inspecting the shadowed crevices of ragged bark, and hopping and flitting among the branches. Sometimes they dangled upside down from the tips of twigs. Because the fires of life needed fuel, the hungry flock soon melted away, and I was alone again with the palpable silence.

In North America, three species of chickadees, also called *tits* (short for titmice), take on

By sunrise, after a long winter night, the energy reserves of a chickadee are nearly depleted.

the boreal forest: the widespread and familiar black-capped chickadee (*Poecile atricapillus*), the boreal chickadee (*Poecile hudsonicus*), which has a brown-colored cap, and the rare Siberian tit (*Parus cinctus*), from Alaska and northern Yukon. In the boreal forests of Eurasia there are two main species of tits: the willow tit (*Parus montanus*), which closely resembles the black-capped chickadee and occupies the same deciduous-forest niche, and the widespread Siberian tit, which is a coniferous forest specialist, as is the boreal chickadee in North America. Chickadees, weighing between a lean-and-mean 8 grams (0.3 oz.) and a meaty 14 grams (0.5 oz.), are the smallest birds to overwinter in the taiga.

This black-capped chickadee's tail was probably bent while the bird spent the night squeezed inside a tree cavity.

Since small birds and mammals are particularly vulnerable to the cold and become chilled far more easily than larger species, how do these tiny tits keep from freezing to death during a taiga winter?

The plumage on a chickadee is much denser than it is on any migrant songbird of the same size. The soft feathers of a black-capped chickadee, for example, can comprise 11.5 percent of its body weight. But the chickadee pays a price for its thick insulating coat of feathers: the tiny bird is a notoriously weak flyer. Tits can further warm up by fluffing out their plumage. The raised feathers trap more air, which improves

their insulating value by a further 30 to 50 percent. When a tit is roosting, it buries its face in its shoulder feathers to lessen the loss of body heat through its eyes and beak, typically areas of high heat loss. For the same reason, it sleeps at night on one needle-thin leg at a time, tucking the other into the warmth of its belly feathers.

In northern Finland, in early January, the long hours of darkness limit the resident tits to just five hours of foraging a day. During the winter solstice in Fairbanks, Alaska, at 64° North, the chickadees there have only three and one-half hours to find food each day. And find food they must, for a chickadee cannot survive a single day without feeding. By nightfall the tiny bird has only enough fat to survive until dawn. Thus, any tactic a tit can use to slow the nocturnal drain on its slim energy reserves may save its life.

Roosting in a sheltered environment is one way tits lessen overnight energy costs. All chickadees typically roost in thick conifers, sheltered from the wind and the thermal drain of the night sky. They may also roost in small tree cavities where there is just enough room for the bird to squeeze inside. Last winter, I spotted a comical-looking black-capped chickadee with the tip of its tail bent to the side like a hockey stick—the likely sequela of a crammed night in a tree cavity.

Willow tits in Sweden sometimes use snow cavities as shelter at night. A determined tit can dig a 20-centimeter (8-in.) tunnel, using its feet, wings, and beak, in less than fifteen seconds. Until recently, chickadees in North America were not known to roost in snow. Then in 1997, researcher Dr. Colleen Cassady St. Clair reported that in northern Alberta, black-capped chickadees sometimes seek refuge in snow holes. She discovered this by attaching tiny radio transmitters, weighing just 0.5 grams (0.02 oz.), to the chickadees and then tracking them to their nocturnal hideaways. Apparently, the tiny tits roost in the snow only on the coldest nights of winter. Dr. Cassady St. Clair speculated that the birds resort to snow roosts when the threat of freezing justifies the risk of being eaten by a keen-nosed predator, such as a weasel (*Mustela*

spp.), fox (*Vulpes vulpes*), or marten (*Martes americana*), that might find them buried in the snow.

For tits, the ultimate strategy to get through the night is to chill out—literally. Birds normally maintain a body temperature of 40°C (104°F), but it is costly for them to keep the heat turned up that high. To conserve energy, many birds lower their internal thermostats at night. The smaller the bird, the more energy it can save by cooling down and the quicker it is able to warm up again next morning. The nighttime body temperature of a roosting Siberian tit, willow tit, or black-capped chickadee may drop to 30°C (86°F), earning the bird an energy savings of up to 45 percent.

All of the taiga tits live where food is abundant in some seasons and scarce in others. They cope with this uneven supply by collecting surplus food in the fat times of autumn and hoarding it for the lean times of winter. During September and October, tits often forage and store extraordinary numbers of seeds, berries, and hibernating adult insects, pupae, and egg clusters. In North America, black-capped chickadees cache hundreds to thousands of food items per day. In Norway, researchers estimated that willow tits may hoard 50,000 to 80,000 spruce (*Picea* spp.) seeds each autumn. And in the Murmansk area of northwestern Russia, Siberian tits stashed away 170,000 edible items from August to November and an additional 30,000 items from December to February.

Chickadees store each food item in a separate site. Common hiding places include lichens, needle clusters, bark crevices, curled leaves, and the broken ends of branches. A quarter of the spruce seeds stored by Alaskan boreal chickadees were also secured with strands of silk scavenged from spider webs and insect cocoons.

The open spruce forest at the tree line offers small birds, such as chickadees, little shelter from the wind.

It is common for a flock of chickadees to mob a roosting saw-whet owl and drive it away.

In the same Alaskan study, researchers watched the chickadees sometimes camouflage the stored food with a bark chip, some lichen, or a spruce needle.

Biologists call the chickadees' style of caching "scatter hoarding." By hiding the food in a multitude of locations, the owner greatly lessens the risk of theft. But, and this is a big but, the hoarder must remember hundreds, if not thousands, of storage sites afterwards. Today, some of the most exciting research in avian memory involves studies with chickadees. The hippocampus is the portion of the brain devoted, in part, to spatial memory. It's no sur-prise that this section of the brain is larger in birds that cache, such as chickadees, than in those that don't. What is surprising, however, is that the size of the hippocampus actually increases and decreases seasonally. In a 1994 study, researchers found that the hippocampus of the black-capped chickadee showed the great-est increase in size in October, the precise time when the bird is caching and needs an improved spatial memory. Noted author and behavioral ecologist Dr. Bernd Heinrich concludes, "It now seems that birds can grow and shrink brain tis-sue as needed, thereby avoiding expensive main-tenance of tissue not in use."

The Hare and
the Hare-chasers

In 1911, legendary naturalist Ernest Thompson Seton described a famous taiga relationship. "The lynx lives on rabbits, follows the rabbits, thinks rabbits, tastes like rabbits, increases with them, and on their failure, dies of starvation in the unrabbited woods." What Seton called a rabbit, today we call a snowshoe hare (*Lepus americanus*), one of the most widespread mammals in the North American taiga. A close relative, the mountain hare (*L. timidus*), hippity hops across the Eurasian taiga. Each continent also has its own lynx with a hunger for hares. In North America it's the 11-kilogram (24-lb.) Canadian lynx (*Lynx canadensis*), and in Russia and Scandinavia, it's the heavyweight Eurasian lynx (*L. lynx*), which can weigh 50 percent more.

Many consider the Russian researcher A. N. Formozov to be one of the fathers of winter ecology. In 1946, he wrote a pioneering paper on the impact of snow on the lives of boreal mammals and birds. Formozov classified wildlife into three categories based on their adaptations to snow. *Chionophobes* were the creatures that had no special adaptations to snow. He included in this group the Eurasian wildcat (*Felis silvestris*), as well as most of the small migrant songbirds such as warblers (Family Parulidae), flycatchers (Order Passeriformes), and sparrows (Family Emberizidae). For them, snow was a hazard to be avoided. *Chioneuphores* tolerated the snow, either by living beneath it, in the manner of voles (*Microtus* spp.) and shrews (*Sorex* spp., *Blarina* spp.), wading through it as moose (*Alces alces*) do, or laboring on top of it as do wolverines (*Gulo gulo*), wolves (*Canis lupus*), and red foxes (*Vulpes vulpes*) do. *Chionophiles* were the snow lovers, the ones with obvious adaptations. These animals turned white in winter and had furry or feathered feet that kept them afloat on the powdery surface of the snow. Formozov considered the willow ptarmigan (*Lagopus lago-*

pus), the arctic fox (*Alopex lagopus*), and the mountain hare to be classic examples of chionophiles.

Boreal hares are well adapted to life in the snow. Beginning in October, the animals replace their brown summer coat with a pure white one, accented with black tips on their ears. In the taiga, the number of hours of daylight, known as the photoperiod, is the most important synchronizer of biological events. For the hare, the shortening days of autumn signal its brain to initiate the complex sequence of hormones that

eventually produces a bleached bunny. Neither temperature nor snowfall affects the timing, which explains why hares may sometimes turn white before it has snowed. Typically, the autumn molt is complete by December.

Researcher Formozov was the first to measure the oversized feet of a hare and appreciate their significance. He calculated the surface area of the animal's four feet, divided the total into its

A male hare is called a buck, the female is a doe, and a newborn is known as a leveret.

In bumper years, hares make up to 93 percent of the diet of the great horned owl.

The great horned owl is the most powerful owl in the North American taiga.

body weight, and obtained an average value of 11 grams per square centimeter (2.5 oz./sq. in.). This value is called the animal's foot loading. The higher its foot loading value, the more likely an animal is to sink into the snow. The foot loading of a red fox, for example, is almost four times greater than it is for a mountain hare, and a wolf's is nine times greater, making the wolf a real sinker. I was curious about my own foot loading, so I traced my size eleven feet on some paper and came up with a value of 217 grams per square centimeter (49 oz/sq. in.). Clearly, I shouldn't be traipsing through the snows of the taiga and should probably stick to asphalt.

No one has yet measured the foot loading of the North American hare; instead, everybody simply quotes Formozov's values for the mountain hare of Eurasia. There is just one problem. The mountain hare can sometimes weigh twice as much as the snowshoe hare, so the foot loading may not be the same for the two species. The true answer awaits the bright-eyed scrutiny of a future graduate student.

Like all hares, the taiga species are built to run. They have long legs, a large heart, ample oxygen-rich myoglobin in their muscles, and a lightweight skeleton. They need speed to outrun the pack of taiga predators intent on eating them. In North America, the red fox, coyote (*Canis latrans*), wolf, fisher (*Martes pennanti*), and marten (*Martes americana*) all dine on *lapin-au-taiga* whenever they can. However, another gang-of-four comprise the real hare-chasers. In bumper years, hares make up to 93 percent of the diet of the great horned owl (*Bubo virginianus*). They are also regularly on the menu of the rapacious northern goshawk (*Accipiter gentilis*), another of the taloned terrors of the boreal forest. Finally, two species of lynx are very closely tied to the lives of taiga hares.

In Formozov's snow adaptation classification, the lynx is a chioneuphore. It endures the snow and makes the best of a difficult situation. With a foot loading value of roughly 35 grams per square centimeter (8 oz./sq. in.), the lynx seems to float over the snow when compared with the

wolf. When compared with the light-footed hare, however, this long-legged feline is somewhat of a lead foot. This may partially explain why the lynx is primarily an ambush hunter of hares and only a part-time snow-stalker and chaser. A study in Alberta showed that once a lynx surprises its prey, a successful hare chase is over in just seven jumps across 12 meters (39 ft.). Even so, the hare outruns the lynx more often than not. On its best days, the cat succeeds in about 40 percent of chases, but usually its success rate is 16 percent or less. In the end, hares comprise about 60 percent of a lynx's winter diet, which also includes red squirrels (*Tamiasciurus hudsonicus*), flying squirrels (*Glaucomys sabrinus*), ruffed grouse (*Bonasa umbellus*), ptarmigan, voles, and deer mice (*Peromyscus maniculatus*). A 1991 paper from Alaska reported on lynx hunts that resulted in some unexpectedly large prey, namely, thirteen successful attacks on red foxes and fourteen cases of predation on caribou (*Rangifer tarandus*) calves. One of the fox kills involved a desperate 200-meter (219-yd.) sprint across the flat, snowy surface of a lake; during one successful caribou hunt, a lynx leaped from a leaning spruce tree and killed the victim by biting its head.

One aspect of the life of the lynx and the hare has received more attention than any other— the regular cycle of boom and bust that each of them follows. Let's begin with the hares. In North America, for the past 150 years at least, the snowshoe hare has been locked into an eight to eleven year cycle of highs and lows. We know this from the fur records kept by the trading posts of the Hudson's Bay Company. During a hare low, you might need to scour 8 hectares (20 acres) of bush to find a single hare, whereas when the hares are happening, you might flush nearly two hundred of them out of the same area. In a typical cycle, if there is such a thing, hares are scarce for three or four years, and then their numbers begin to build. After several years of increase, their population reaches a peak. In Alberta, veteran hare researcher Dr. Lloyd Keith reported peak densities greater than 3,100 hares per square kilometer (8,030/sq. mi.). Peak

A lynx kills an average of two snowshoe hares every three days.

numbers last for a year or two; then the population suddenly crashes, and from 80 to 90 percent of the hares die. In the Russian taiga, mountain hares follow a similar cycle, but in Scandinavia, the hare cycle is much shorter, lasting just three to four years.

The lynx populations on both continents shadow the hare cycle, but usually lag a year or two behind them. When hares are scarce, adult lynx begin to starve, females fail to conceive, or at best, females produce small litters of tiny kit-

tens, most of which subsequently starve. Yearlings are the underdogs of lynx society, and their low rank and inexperience forces them to disperse long distances in search of food. The record journey belongs to a juvenile from Snafu Lake, Yukon, that wandered east to Wood Buffalo National Park, an 1,100-kilometer (684-mi.) trek.

When hare numbers first plummet, lynx may sustain themselves for a year or so by switching prey and hunting grouse and ptarmigan. Eventually, the heavy predation pressure on

Rivers are common travel corridors for a lynx moving about in its winter territory.

these birds also causes them to decline in number. At this point, the lynx population has nowhere to go but down, and their numbers may plummet to a quarter of what they were. For three or four years after this, no kittens survive to adulthood. Once the hares begin to rebound, lynx numbers slowly recover.

For decades, biologists have debated why taiga hare populations cycle in the first place. It seems that cycling may result from a combina-tion of arboreal chemical warfare, predators, and perhaps even sunspots. It's likely that most her-bivores, including hares, select their food based on its caloric value, nutrient content, and the level of chemical deterrents that are present. In winter, snowshoe and mountain hares eat woody vegetation, namely, the bark and twigs of willow (*Salix* spp.), aspen (*Populus tremuloides*), birch (*Betula* spp.), and spruce (*Picea* spp.). As the hare population increases, the browsing pres-

and inhibit the growth of digestive microbes. At the peak of the hare cycle, the animals begin to weaken and die from malnutrition. At the same time, peak populations of lynx and other predators take their toll. The combined assault of chemistry and carnivores defeats the hares, and their population collapses.

In the years that follow, woody plants slowly recover. With fewer hares browsing on them, they have less need for chemical deterrents, and their levels drop accordingly. As plants become less toxic and more digestible, hare numbers slowly begin to rebuild. At the same time, correspondingly low predator populations also benefit the hares.

So where do sunspots fit into this story? Sunspots follow an average cycle of 10.6 years. It may not be coincidence that this is also the average length of the hare cycle. Researcher Dr. Anthony Sinclair at the University of British Columbia believes the two are connected. Perhaps sunspots influence the weather, and the resulting changes in snowfall and precipitation may affect the growth of the young trees upon which the hares feed. The delicious complexity of it all has still to be deciphered.

At the end of winter, a hungry lynx is not the only sprinter likely to chase a hare. It's March madness in bunnyland, and hormones are raging. In the early stages of the breeding season, several male hares may chase a single female. The rival males thump their back feet and chatter their teeth at each other in aggression. The fact that 20 percent of the bucks carry bite scars attests to the seriousness of these threats. In the end, the lucky suitor gets to romp in the snow with the female, in a rather unconventional courtship. As one of the pair leaps in the air, the other one runs beneath it and is sprayed with urine by its airborne partner.

Let's end with a naturalist quickie quiz. How can you tell that the hare-breeding season has begun in the taiga?

The answer: splattered urine is clearly visible in the snow. The telltale yellow stains may cover an area 9 meters (30 ft.) in diameter.

sure on woody plants also increases. The plants react by flooding their tissues with toxic metabolites (tannins, phenolic resins, and terpenes), which makes them less palatable, thus deterring browsers.

Scientists speculate that as this so-called chemical warfare escalates, hares continue to browse the plants, attempting to meet their nutritional needs. However, the ingested toxins reduce a hare's ability to absorb vital nutrients

The Underside of Winter

The brown lemming's globular shape helps reduce body heat loss on cold days.

On the edge of a sphagnum bog, deep in the taiga, a frozen wind is hissing through the needle-bare branches of a tamarack. Tiny microblades of ice skid across the wind-glazed surface of the drifted snow. The sun is in full winter glare, and the temperature is -34°C (-29°F). The wind and ice in the air make it almost too painful to breathe. But a mere 60 centimeters (2 ft.) beneath the webbing on my snowshoes,

there is another world, the subnivean world at the base of the snow. Winter ecologist Dr. Peter Marchand aptly describes this hidden world as "the underside of winter." Here, the temperature wavers around freezing all the time. This is a world of deep blue shadows, since less than 1 percent of daylight reaches this depth. The air is saturated with water vapor, and the dank darkness is rich with the earthy smells of decaying plants and the musky scents of rodents and shrews. The only sound is the occasional tinkle of falling hoarfrost crystals, disturbed by a scuttling spider or beetle. In his classic book, *Life in the Cold: An Introduction to Winter Ecology,* Dr. Marchand enthuses, "Maybe the hardest notion to accept about winter is that it is so alive. Beneath the bark of the leafless tree, under the frozen moss, in all the crevices of winter, there is life!"

For many of the large creatures of the taiga, the snow is a seasonal hardship. It hampers movements, buries food, and depletes energy reserves. For most of the smaller creatures, however, the snow is a godsend, for without it, they could never endure the cold of winter. It's the physical nature of snow and how it changes that makes it a lifesaver for many. Light fluffy snow

is 92 percent air, and even after the snow settles, the proportion of air may still be as high as 70 percent. The air trapped between the crystals of snow makes it a good insulator.

Manitoba researcher Dr. William Pruitt, known affectionately by his colleagues as the "snowman," was first to coin the expression "hiemal threshold." This is the minimum amount of snow cover necessary to insulate the ground and protect the winter creatures that live next to it. The threshold varies from 20 to 30 centimeters (8 to 12 in.), depending upon the degree of snow compaction. In winters when snowfall is late in coming, the frigid taiga temperatures may produce a lethal "freeze out," in which small mammals die, and ground plants are killed.

A fresh snowfall undergoes two changes that affect wildlife. Within hours, the delicate needle-like crystals of snowflakes break, and the fragments pack closer together—the work of wind and gravity. The resulting compacted snow is denser, and although it loses some of its insulation quality, the higher density of the snow makes the surface stronger and thus better able to support the weight of animals walking on top. Snow scientists call this first process "destructive metamorphosis." A second process, called "constructive metamorphosis," is more important for the subnivean world. When a snowfall first covers the ground, heat from the earth warms the thin layer of air sandwiched between the ground and the bottom of the snow. This melts some of the snow and transforms it into a loose latticework of delicate, brittle crystals referred to as "depth hoar." Here, in the crevices and corridors of this crystalline world of frost, an entire ecosystem thrives through the winter.

The subnivean ecosystem, like every ecosystem, is built on a foundation of plant life. Beneath the snow's protective mantle, dead plants continue to decay, and fungi sprout easily in the cool temperatures. Some green plants can grow and photosynthesize even when they are buried under 80 centimeters (31 in.) of snow. A surprising array of creatures feeds on these plants. In Manitoba, researcher Cassie Aitchison winter-trapped sixteen species of tiny, wingless springtails (Order Collembola), as well as minute sap-sucking aphids (Family Aphidae), leafhopper nymphs (Family Cicadellidae), and linidae), fifteen species of predatory and parasitic mites (*Acari* spp.), and nineteen species of spiders, most of which were big-fanged wolf spiders (Family Lycosidae) or predaceous running spiders (Family Clubionidae). A few other spiders (Family Linyphiidae) even built webs in the snow crevices to ensnare springtails. Completing the guild of predators were carnivorous pseudoscorpions (Order Pseudoscorpiones) and venom-clawed centipedes (Class Chilopoda).

As fascinating as the invertebrates are, it is the relatively big critters of the subnivean world that get all the attention. Across the taiga of North America, roughly twenty species of small

Scientists use many names to describe snow: qali *for snow on trees,* api *for snow on the ground, and* pukak *for the crystallized bottom layer of snow.*

numbers of scavenging ground beetles (Family Carabidae). All these insects were active in winter and feeding on rotting plants, fungi, and greenery, as well as on the bodies of dead insects.

For every insect that eats, there is an insect eater, and the underside of winter is a veritable subnivean Serengeti Plains. Researcher Aitchison winter-trapped several dozen kinds of carnivorous rove beetles (Family Staphy- mammals overwinter under the snow, and none of them hibernates. They include six species of common voles (*Microtus* spp.), two species of red-backed voles (*Clethrionomys* spp.), the eastern heather vole (*Phenacomys ungava*), six species of shrews (*Sorex* spp., *Blarina* spp.), two species of bog lemmings (*Synaptomys* spp.), the brown lemming (*Lemmus sibiricus*), the starnosed mole (*Condylura cristata*), and the deer mouse (*Peromyscus maniculatus*). The members of this squad of small mammals adopt one of two lifestyles to weather the winter—they either huddle together and socialize, or they live as shrunken loners.

The voles, mice, and lemmings are primarily vegetarians. They gnaw on roots, bark, shoots, twigs, and fungi, with an occasional side order of insects. To reach their food, the rodents excavate a network of runways under the snow and also dig tunnels through the middle layers of the snow pack, some of which connect to the surface. In Finland, researcher K. Korhenen counted up to eighteen air vents in 100 square meters (120 sq. yd.) of snow surface. Last winter, in northern Alberta, I counted four such air vents in 1 square meter (1.2 sq. yd.) of powder snow. Formozov was the first to propose that voles excavate these air holes to ventilate the deeper parts of their tunnel system, thus adjusting oxygen and carbon dioxide levels. However, this belief has become part of taiga folklore. Scientists have repeatedly analyzed the oxygen and carbon dioxide levels in subnivean tunnels, but they have never been able to prove that these air vents alter the gas composition in any favorable way. In the end, the reason why voles dig these air vents is still a mystery.

Even though the temperatures within the subnivean world are moderated by the snow cover, they are still cold enough to challenge small rodents. All of the voles, commonly called field mice, have a globular body shape, short legs and tail, and small ears, shapes that minimize their surface area and help them conserve body heat. If small mammals were to insulate themselves with thick layers of fat or fur, their mobility would be seriously impaired. Instead, they build insulated nests of grass and huddle together with their neighbors. As many as twenty-eight deer mice have been found sharing a single winter nest, and it's not uncommon to find from six to ten voles huddled together in a nest. For this to happen, the rodents undergo a dramatic personality change in advance of winter.

Throughout the summer breeding season, all voles and mice are solitary and fiercely territorial, defending their personal plot of turf with tooth and nail. However, with the onset of autumn, they suddenly become sociable and docile and remain this way throughout winter. The thermal advantages of huddling apparently prompt the animals to transform from summer samurai to winter nest mates.

The subnivean shrews have their own remarkable way of dealing with winter. Shrews are the tiniest mammals in the boreal forest. The smallest shrew of all is the widespread pygmy shrew (*Sorex hoyi*) of North America. It weighs a mere 5 grams (0.2 oz.), no more than a quarter, and it takes ten of these miniature insectivores to equal the weight of a single meadow vole (*Microtus pennsylvanicus*). For much of the year, shrews are loners. Each scours its own tiny territory, attacking anything that scuttles, scurries, or slithers. It hunts insects, spiders, mites, and centipedes, although it also eats conifer seeds. A shrew has a truly voracious appetite and eats every two to three hours, day and night, consuming a daily weight in prey equal to 80 or 90 percent of its own body weight.

Shrews live life in the fast lane, but they burn out quickly. British writer Sara Churchfield explains in her fascinating book *The Natural History of Shrews* that "most northern shrews are annuals. Following birth in the summer, the shrews overwinter as immatures, breed in the following spring, then die ... usually by September."

The boreal shrews have come up with a truly novel way to endure winter under the snow— they literally shrivel up. In preparation for winter, a shrew may lose up to 45 percent of its body weight. It doesn't simply lose body fat. The animal's skull shrinks, its backbone shortens, its muscles thin out, and its liver, kidneys, and spleen become smaller. Most remarkable of all is that the animal's brain shrivels by up to a third. By reducing the weight of these vital tissues, the shrew greatly reduces its daily energy needs and improves the chances it has of surviving until spring, earning the one opportunity it will ever have to pass its genes to the next generation.

A red fox in northern Saskatchewan tries to catch the vole it hears moving under the snow pack.

For the large creatures of the taiga, snow is a seasonal hardship. For most of the smaller creatures, however, the snow is a godsend, for without it, they could never endure the cold of winter.

River otters, like all boreal mustelids, pair with a mate for a few days during the breeding season and then separate immediately afterwards.

The taiga of North America is home to eight species of mustelids, and a dozen or so live in the boreal forests of Eurasia.

The Weasel Clan

Worldwide, there are roughly 4,450 species of mammals, and 240 of these belong to the Order Carnivora. Among the carnivores, the most varied group are the mustelids, or weasel family (Family Mustelidae), consisting of about sixty-five species. The mustelids are small to medium-sized predators, many of which have lithe, slender bodies and short legs. Most are capable hunters with powerful jaws, sharp canines, and a capacity to kill prey much larger than themselves with a lethal bite to the nape of the victim's neck. All of the mustelids stink, or at least they do to the sanitized nose of modern humans. The animals have well-developed anal glands that produce a thick, musky oil that they mix with their feces and leave along trails and other conspicuous sites to advertise their land claims. The smelly anal secretions can also be discharged when the animal is frightened, and no mustelid does this with more panache than the striped skunk (*Mephitis mephitis*). Ernest Thompson Seton described the nauseating skunk smell as "a mixture of strong ammonia, essence of garlic, burning of sulphur, a volume of sewer gas ... and a dash of perfume musk."

The mustelids are the most diverse carnivores in the boreal forest. The taiga of North America is home to eight species, and a dozen or so live in the boreal forests of Eurasia. Such diversity is uncommon in a community of carnivores, and it's interesting to examine how the North American mustelids masterfully partition the taiga to exploit every possible feeding opportunity.

The circumboreal least weasel (*Mustela nivalis*) is the smallest carnivore in the world. Weighing just 50 grams (1.75 oz.), this miniature mustelid can squeeze through a wedding ring. It and the larger short-tailed weasel (*M. erminea*) are the vole specialists of the boreal. Their long, slender bodies allow them to tunnel beneath the snow, penetrate the subnivean world, and pursue prey along its runways. These hyperactive carnivores always move at full speed. They don't run so much as flow from spot to spot, checking every hole and crack as they go.

To lessen the competition between them, the least and short-tailed weasels commonly concentrate on different-sized prey and hunt in different areas. The diminutive least weasel favors red-backed voles (*Clethrionomys* spp.) and deer mice (*Peromyscus maniculatus*), which it finds in forested areas, whereas the heavier, short-tailed weasel tackles the large meadow voles (*Microtus pennsylvanicus*) and bog lemmings (*Synaptomys* spp.) that live in willow (*Salix* spp.) and alder (*Alnus* spp.) thickets and sphagnum bogs.

The two weasels are the only taiga mustelids that turn white in winter. Though they often hunt beneath the snow, both of these small carnivores travel on top of its surface when they are moving about their territories. The winter color change is a tactic that makes them less conspicuous to hungry predators, rather than camouflaging them from their prey.

The American marten (*Martes americana*) and the fisher (*M. pennanti*) are the arboreal mustelids of the taiga. These agile carnivores can race among the branches of conifers with astonishing speed and fluidity, counterbalanced by their long bushy tail. Well-honed for life in the trees, they both have rear feet that rotate backward so that they can climb down headfirst. As adept as they are in the trees, the marten and fisher do most of their hunting on the ground, where they prey on hares (*Lepus* spp.), grouse (*Falcipennis* spp., *Bonasa* spp), voles, and mice. In fact, the Cree name for the marten is *wabachis*, which means hare-chaser. The 2.3- to 4.5-kilogram (5 to 10-lb.) fisher, four times the weight of a marten, preys on its smaller relative whenever it can catch it. The fisher's greatest claim to fame, however, comes from its ability to successfully prey on the porcupine (*Erethizon dorsatum*), something that no other taiga

The noxious smell of the striped skunk does not deter the great horned owl, a common skunk predator.

predator does with such skill and regularity.

The back of a porcupine's neck is protected by a thick mat of sharp quills, so the fisher cannot attack this area as it does with most other prey. Instead, it repeatedly attacks and bites the animal's unquilled face. Although it may take half an hour or more, the porcupine gradually weakens and eventually succumbs. The killer then eats the victim through its belly, where there are no quills. Despite taking such precautions, as many as 18 percent of fishers have old quills imbedded in their body.

The taiga is as much a waterscape as it is a forested landscape. The myriad lakes, beaver ponds, rivers, and streams are the hunting grounds of two other mustelids, the semi-aquatic mink (*Mustela vison*) and the water-loving river otter (*Lontra canadensis*). The mink is a shoreline predator, hunting small rodents and birds on the edge of the water and crustaceans, frogs, and small fish, such as sticklebacks (Family Gasterosteidae), in the shallows. Although mink are strong swimmers, they are not well adapted to hunting underwater. Their eyesight is designed for above water, their feet are only partially webbed, and they can only stay underwater for about twenty seconds. One cold afternoon in February, I watched a hungry mink eat three adult leopard frogs (*Rana pipiens*). The animal dove to the bottom of a shallow stream and discovered the amphibians hibernating in the rubble. It brought the frogs to the surface, one at a time, and ate them at a leisurely pace as it sat on the edge of the ice.

The river otter is a specialized aquatic predator. It has a thick, waterproof coat of fur, fully webbed feet, and a laterally flattened muscular tail. The otter hunts in deeper water than mink do and can stay submerged for up to four minutes while it searches for fish, its principal diet. Despite its underwater prowess, the otter frequently targets sluggish fish such as perch

(Family Percidae), sunfish (Family Centrarchidae), and suckers (Family Catostomidae) rather than fast-finning trout (Family Salmonidae), pike (Family Esocidae), and pickerel (*Esox* spp.).

Winter in the boreal forest can be taxing when you're an aquatic predator because the lakes are covered with up to a meter of ice. A biologist friend, Dr. Don Reid, discovered how otters improve their winter fishing conditions. The enterprising mustelids sometimes break through a beaver dam to lower the water level. This concentrates the fish population and creates a vital air space under the ice. It also allows the otters to access adjoining ponds without exposing themselves to the weather. Since beavers never repair their dams in winter, the renovations benefit otters until spring.

Two other mustelids, the striped skunk and the wolverine (*Gulo gulo*), have lifestyles quite different from the rest of the taiga group. The skunk is a digging omnivore, and the wolverine is a bone-breaking scavenger. The skunk is most at home in open areas, especially around human habitations, and avoids unbroken tracts of forest. This familiar weasel eats grasshoppers, berries, and birds' eggs. It overturns rocks to search for beetles and uses its 2.5-centimeter (1-in.) front claws to dig for grubs. All of these foods, as you can imagine, become scarce or impossible to find once the snow flies. To avoid the demands of winter, the skunk dens from November to March, something no other taiga mustelid does. A single adult male skunk may den with five or six females, or as many as nineteen, as was the case in one Alberta den. Once inside the den, the animals' body temperature may drop by as much as 10 C° (18 F°), an important strategy for conserving energy.

It takes a special breed of scientist to work with skunks. Apparently you can safely handle the stinkers without being bitten or sprayed if you simply grab the animal around the neck with one hand and hold its tail with the other. The authors of this sage advice suggest, "the animal should then be turned on its back with the tail pointed away and downwind." No kidding.

The much maligned, but rarely seen, wolverine has a number of other common names, none of them complimentary: glutton, skunk bear, and devil bear. The adult male wolverine, weighing up to 27 kilograms (60 lb.), resembles a small bear with a bushy tail.

If you examine the skull and teeth of any mammal, you can learn a lot about the owner's lifestyle and diet. For the size of the animal, the skull of a wolverine is massive and broad. It has well-developed anchor points for its powerful jaw muscles and strong neck muscles, which it uses for lifting and dragging. Its molar teeth are large and solid, designed to cut tough cartilage and ligaments, and crush all but the largest of bones. Obviously the wolverine's impressive dentition and skull anatomy did not evolve to pluck berries and capture small mammals; they are the tools of a major scavenger.

When conditions are right, the wolverine is also a capable predator. It has been known to tackle and kill large mammals, such as moose (*Alces alces*), caribou (*Rangifer tarandus*), and reindeer (*Rangifer tarandus tarandus*). These, of course, were exceptional cases, in which the victims were usually mired in deep snow, giving the wolverine a killing advantage. The foot loading of a wolverine is the same as that of a lynx (*Lynx canadensis*), so the animal can run easily on the snow's surface.

In an Alaska study, Audrey Magoun followed 80 kilometers (50 mi.) of wolverine winter trails, attempting to discover the dietary details of this mysterious animal. She found 186 "digs" along the trails. Among these, sixteen digs contained ground squirrel remains, six contained flecks of blood, probably from live voles or lemmings, sixteen contained splinters of caribou bone, five contained ptarmigan feathers, one contained a whole shrew, one contained the dried, mud-caked carcass of a duck, and three contained eggshells. There was no evidence of food remains in the rest of the digs. Magoun concluded, "The wolverine's ability to survive the most severe time of the year, on such a meagre diet, attests to its efficiency as a scavenger."

The fascinating sex life of the mustelid clan

The American marten, which is the size of a small house cat, hunts voles, deer mice, squirrels, hares, and grouse.

deserves a few parting words. In every case, male mustelids are heavier than their mates—from 10 to 20 percent heavier among skunks and river otters, and up to 100 percent heavier, or more, among fishers, short-tailed weasels, and wolverines. The most common explanation for such size differences, called *sexual dimorphism*, is competition among rival males for access to females in heat.

To produce the strongest offspring, females need to mate with the healthiest male available, not just any big male that comes along. A small female cannot prevent a male from mating with her without starting a fight in which she might be seriously injured by her stronger partner. Her solution is to mate with whichever male corners her and then to let her internal physiology determine whether he has the "right stuff" to sire her offspring. How does she do this? By induced ovulation! Most mustelids are marathon copulators, mating for hours at a time and repeating this strenuous feat numerous times over several days. Somehow, the genital area of a female mustelid evaluates the vigor of a male's copulatory efforts, and if they fall short of a predetermined target level, the female fails to ovulate. Only a healthy, virile male can sustain the frequency and vigor necessary to induce ovulation. In the world of mustelids, the females have the final word.

Hammerheads

When I was ten years old, my uncle gave me a bird book illustrated with paintings by the famous artist John James Audubon. There were at most a hundred birds pictured in the book, and I was certain they included every bird in the world (the world total is actually over 9,800 species). For over a year I tried, unsuccessfully, to match the birds I saw near my home in Ontario with the ones illustrated in the book. What I didn't know at the time was that most of the paintings were of Florida birds, a sure way to frustrate a nascent nature nerd. I finally got lucky one winter afternoon while I was skating along the frozen shoreline of a lake. I flushed a large bird from a pine tree (*Pinus* spp.), and it squawked, took off, and flew low over my head. The mysterious bird was as big as a crow, and had an inky-black body, large white patches on its wings, and a bright red crest. Now this was a bird I was certain I could identify, and sure enough, there it was on page 56 of my bird book. It was a pileated woodpecker (*Dryocopus pileatus*), the largest woodpecker in North America and the inspiration for the cartoon character Woody Woodpecker. Forty-two years later, I still remember that sighting as a magical moment of discovery.

In a simplistic way (which is the only way I understand botany), the boreal forest is composed of just three kinds of vegetation: trees, trees, and more trees. In every forest of trees there are columns of tree trunks, sometimes more than 5,000 trunks per hectare (2,024/acre) of forest. Even so, few birds have successfully exploited the feeding possibilities of these trunks. For example, among the several hundred species of songbirds that visit the taiga, only four species regularly forage on tree trunks: the two

The crow-sized pileated woodpecker may roost in half a dozen different tree cavities during a single winter.

The northern flying squirrel nests in abandoned tree cavities excavated by pileated wood-peckers.

nuthatches (*Sitta* spp.), the brown creeper (*Certhia americana*), and the black-and-white warbler (*Mniotilta varia*). Enter the woodpeckers. Exquisitively skilled to tackle a tree trunk, they hammer, chisel, tear, and probe better than all other birds and are a conspicuous presence in the winter taiga. In the boreal forests of North America there are seven species of woodpeckers (six species in Eurasia), ranging in size from the small downy woodpecker (*Picoides pubescens*), which weighs less than a fork full of butter pecan pie, all the way up to the 42-centimeter (16.5-in.) pileated woodpecker, which could eat the whole pie.

Woodpeckers have cornered the market on the tree trunk habitat because of a unique set of adaptations. The usual toe arrangement in songbirds consists of three facing forward and one facing backward. This may be ideal for perching on a horizontal branch, but it's ill designed for climbing a vertical trunk. The foot of a woodpecker has two toes facing forward and two facing backward, which produces a more secure grip on the bark. A woodpecker also has very strong thigh muscles to hold itself firmly against the trunk, and the shafts of its central tail feathers are reinforced and pointed, to better serve as a prop against the bark. It's no coincidence that all boreal woodpeckers have black-tipped tails. The black color results from the presence of melanin, a durable pigment that makes the feather tips more resistant to wear and abrasion.

As you might expect in a chronic head banger, the skull of the woodpecker features some interesting adaptations. In 1976 the prestigious British medical journal *Lancet* addressed this important topic in a technical paper entitled "Woodpeckers and Head Injury." The four authors noted that a woodpecker may hammer its head five or six hundred times a day and they wondered "why the countryside was not littered with dazed and dying woodpeckers?" Ultimately, the neuroscientists could not explain a hammerhead's apparent immunity to head injury. Perhaps the bird's small brain, cushioned by a thin layer of spinal fluid, fails to transmit dangerous shock waves, or perhaps certain muscles at the base of the bird's jaws function like shock absorbers to dampen the force of the impact. Still, no one knows for sure, and the hardheaded carpenters hammer on.

The tongue of a woodpecker is another avian believe-it-or-not. In most species of birds, the

tongue ends at the back of the throat. This is not so in the hammerheads. In many species, including the yellow-bellied sapsucker (*Sphyrapicus varius*), the tongue muscle extends around both sides of the neck and up the back of the skull to the top of the head, where the two parts reunite. The length of this muscular attachment determines how far the bird can extend the sticky tip of its tongue

An ant-eating bird such as the northern flicker is called a formicivore.

The yellow-bellied sapsucker leaves the taiga in autumn and migrates to the southern United States.

beyond the end of its beak. In the hairy woodpecker (*Picoides villosus*), the bird's tongue muscle loops over its head and then curls into the bird's orbit and attaches behind its eyeball. The northern flicker (*Colaptes auratus*) has the longest tongue of all. After tracing over the top of the bird's head, the tongue passes through the flicker's nostril and attaches at the end of its beak. The long attachment enables the flicker to stick its tongue out nearly 5 centimeters (2 in.) beyond the tip of its bill.

Author and scientist Dr. Robert McFarlane describes the action of the woodpecker's tongue in his fascinating book *A Stillness in the Pines: The Ecology of the Red-Cockaded Woodpecker*.

The tongue of the woodpecker has adapted into a specialized extensible organ of fantastic dexterity. It can protrude beyond the tip of the bill several inches in some species and terminates in a hard, horny tip armed with minute spines. Once a woodpecker penetrates the tunnel of a wood-boring insect larva, it can explore the tunnel with its tongue, pierce the soft body of the larva with the sharp horny tip and extract the prey impaled on the spines. Large salivary glands produce sticky mucus that lubricates the tongue to allow smooth passage through insect tunnels and snag ants and other insects on the surface of the tongue.

Five of the seven taiga woodpeckers are year-round residents. Only the sapsucker and the northern flicker migrate to the southern United States for the winter. All of the overwintering woodpeckers search for food in four ways: they hammer and listen for grubs, search and probe crevices and cracks, flake off bits of bark to expose beetle tunnels or their builders, and glean needle clusters for insect hiding places. Ants are the most common food for all woodpeckers, but in winter the birds also hunt hibernating spiders and insects, as well as their eggs and pupae. As an example, coccids (*Xylocculus betulae*) are tiny sap-sucking insects that embed themselves in the bark of paper birch trees (*Betula papyrifera*). On a single tree there may be hundreds of coccids, which make a sweet winter meal for a downy woodpecker.

Since the taiga woodpeckers forage in similar ways, and often eat the same foods, how do they avoid competing with each other? They do it by exploiting slightly different areas. Some woodpeckers feed on the ground; others feed on the trunk, either within the crown or below the crown; still others forage on primary or secondary limbs. Conifers, deciduous trees, dead trees, burned trees, and live trees are also partitioned. The greatest competition actually comes from a woodpecker's own mate, who likely feeds in a very similar fashion. Even in this situation, the pair may partition their feeding areas. For example, among downy woodpeckers, males frequently feed on the upper trunk, limbs, and smaller branches of paper birch, whereas the female concentrates on the middle and lower regions of the trunk.

One of the unique characteristics of all woodpeckers is their use of drumming to communicate. Everyone has heard the characteristic hollow hammering on a tree snag or branch, and if you are like me, you assumed it was a pumped-up male advertising his virility to rival males and prospective female mates. Well, it turns out that woodpeckers hammer their hearts out for this and many other reasons, and females may drum as often as males. In winter, a pair of woodpeckers will usually roost in separate tree cavities overnight, to shelter from the weather. At dawn, mates may drum to each other to coordinate a rendezvous. Both sexes may also drum when an intruder enters their territory, as a displacement behavior meant to work off anxiety. When woodpeckers are feeling relaxed and content, they may drum between preening bouts. Both sexes may also drum to notify their partner that they wish to mate. It's a little like Tarzan beating his chest. One January in northern Saskatchewan, I saw a pair of hairy woodpeckers drumming for yet another reason. The drumming of one woodpecker stimulated a burst from its partner. The duet lasted nearly ten minutes, and the two birds each drummed at least half a dozen times.

Common golden-eyes, such as this courting male, nest in old woodpecker holes.

Hairy woodpeckers pair in midwinter, and the drumming duet was probably a means of reinforcing the pair bond during the three or four months before nesting.

For a human, the drum roll of one woodpecker is hard to distinguish from that of another. But can the woodpeckers differentiate? You bet they can. A California researcher, Dr. Eric Johnson, and his students recorded the drums of eleven different species of woodpeckers and then analyzed them. They discovered that drumming sequences are unique for each species; they differ in the number of beats per second and also in whether the drum sequence speeds up, slows down, or remains steady. Here are their findings for some of the boreal woodpeckers. Pileated woodpeckers drum the slowest at fourteen beats per second and speed up during a sequence; black-backed woodpeckers drum at sixteen beats per second and also speed up; downy woodpeckers tap at seventeen beats per second and slow down toward the end of a sequence; northern flickers rap at a steady pace, at twenty-two beats per second; hairy woodpeckers drum at twenty-six beats per second, also at a steady pace; and three-toed woodpeckers drum the fastest at twenty-eight beats per second, either at a steady pace, or slowing down toward the end of the sequence. Another interesting observation is that when two species of woodpecker overlap in their feeding behavior or habitat preferences, for example, downies and hairies, or black-backs and three-toeds, the rate of their drumming is vastly different. In fact, they are at opposite ends of the scale, so there can be no mistaking their identity. Such subtle communication methods demonstrate that woodpeckers aren't boneheads after all.

Caribou—The Snow Deer

Between scattered copses of stunted spruce, the ground is covered with 15 centimeters (6 in.) of powdery snow. It's October along the taiga's northern tree line. The days are bright and cold, the nights clear and colder. The air is filled with promise—the caribou are in rut. Heavy-antlered bulls with muscled necks chase rivals to exhaustion, their tongues lolling and their bellies heaving. Nervous cows mill around the battling bulls, each hoping to be overtaken by one of the victors. The fighting and the coupling is over in a matter of days. Eighty to 90 percent of the cows will mate during a ten-day period in the middle of the month, ensuring that their calves are born in synchrony the following June, thus glutting their predators with excess. The cows are at their fattest this time of the year, the bulls, at their leanest. Ahead of all caribou, there is half a year of snow and cold. It is now the season to conserve and survive.

Perhaps two and a half million restless caribou (*Rangifer tarandus*) range across the arctic tundra of mainland North America. In the mountains of Alaska and Yukon there are Grant's caribou (*R. t. granti*). Across the rolling hinterlands of the Northwest Territories and Nunavut there are the barrenground caribou (*R. t. groenlandicus*). Finally, in northern Quebec and Labrador there is the greatest aggregation of all—the 800,000-strong George River Herd of woodland caribou (*R. t. caribou*). Even though some of these hardy animals tough out the win-

Typically, the taiga wintering areas of caribou include open spaces where the animals can forage beneath the snow.

ter blizzards and compacted snow of the arctic tundra, most stream south in autumn. They cross the taiga tree line in waves and penetrate up to 320 kilometers (200 mi.) or more into the northern boreal forest. Here they may encounter the resident woodland caribou that sparsely inhabit the forests of the western and central taiga. Throughout winter, caribou wander in small bands comprising up to two dozen members that search for open conifer forests, large frozen lakes, and snow with just the right depth and hardness. Caribou are the snow deer, better adapted to it than any other hoofed mammal on Earth. The taiga is their winter world, and they have learned to master it.

Imagine enduring a temperature range of 83 C° (150 F°). Begin on the warmest day in summer, at 28°C (82°F), when blood-sucking mosquitoes, as thick as smoke, swarm around your eyes. Then let the temperature fall past the cool days of late summer, through the frost and freezing cold of autumn, to the metal-shattering frigidity of winter at -55°C (-67°F). This is the annual life of the caribou. Its thick winter coat is 5 centimeters (2 in.) long, and the hollow, tapered hairs provide remarkable insulation against the cold. Only when the temperature sinks lower than this does the animal need to increase its metabolism to stay warm. A caribou's broad hooves and prominent dewclaws, the two small protuberances at the rear of each foot, result in another winter advantage—they provide a large surface area to support its weight on the snow.

The caribou's behavior, as much as its anatomy, is adapted for snow. Caribou are inveterate followers. Well-worn trails lessen the energy required by an animal to move from place to place. When they can, caribou also travel on frozen lakes, where the snow is compacted by the wind and the walking is easier. They also use these lakes to rest and ruminate. Good visibility permits earlier detection of wolves, and the hard snow helps the caribou to outrun its enemies. For these reasons, taiga wintering areas of caribou are generally rich with lakes.

Caribou rarely browse on woody vegetation

The autumn rut of many barren-ground caribou occurs in the open forests along the northern fringe of the taiga.

When a grizzly first leaves its winter den, it scavenges any dead caribou that succumbed to the rigors of the season.

above the snow, which is the way moose feed during winter. Instead, caribou use their broad blunt snout to graze vegetation on the ground. To reach their food, they have become experts at digging, something that no other deer does with such skill. Caribou use their sharp-edged front hooves to dig as many as 140 snow craters in a day. The animal's name is derived from the Micmac Native word *xalibu*, which means the "pawer." In the days of myth and folklore, people believed that caribou used the broad brow tine of their antlers to dig through the snow, which is why it is called the shovel. Alas, this just ain't so. In fact, caribou bulls, which have the largest shovels, shed their antlers in early November, before the deep snow even arrives.

The winter diet of the caribou includes a mixture of three foods: lichens, sedges (*Carex* spp.), and the evergreen leaves of crowberry (*Empetrum nigrum*), bearberry (*Arctostaphylos uva-ursi*), labrador tea (*Ledum groenlandicum*), and leatherleaf (*Chamaedaphne calyculata*). Lichens, the so-called reindeer mosses of the genus *Cladonia,* may constitute half the animals' diet. At least twenty different varieties of *Cladonia* grow in the taiga of western North America alone. The caribou's taste for lichens is unique among hoofed mammals, but the animal cannot live on lichens alone. If they eat nothing else, caribou lose weight, even if they consume 5.4 kilograms (12 lb.) of the crunchy stuff a day, which is theoretically enough to meet their daily caloric needs. To stay healthy, a wintering caribou must eat other plants as well.

Snow-digging caribou are very sensitive to the depth and hardness of the snow. In northern Manitoba, Dr. William Pruitt calculated that the optimal snow depth for caribou was less than 60 centimeters (24 in.) and the critical hardness was 50 grams per square centimeter (11 oz./sq. in.). When either of these parameters was exceeded, Dr. Pruitt could predict that the caribou would move to a new area of forest, in search of easier digging conditions. Even a single storm can affect caribou movements. After a two-day storm in northern Saskatchewan, the snow hardness doubled, and the caribou moved away. In Alaska, caribou often contend with much harder snow, which sometimes reaches a hardness level of 6,500 grams per square centimeter (1,468 oz./sq. in.). In Labrador, the snow conditions can be even more demanding. Winter snowfall sometimes exceeds 400 centimeters (157 in.), and freezing rain can transform the snow into veritable concrete with an impenetrable hardness as high as 50,000 grams per square centimeter (11,290 oz./sq. in.). At times like these, there should be some sort of government affirmative action plan to issue jackhammers to hungry caribou.

When the snow is too deep or impossible to dig, caribou switch their diet to tree lichens. In northern Saskatchewan, researcher Dr. George Scotter considered horsehair (*Bryoria* spp.) and old man's beard (*Usnea* spp.) to be important lichens for wintering caribou. There are many kinds of usnea tree lichens, some of which have delightful common names, such as scruffy old man's beard (*U. scabrata*), powdery old man's beard (*U. lapponica*), shaggy old man's beard (*U. hirta*), and pitted old man's beard (*U. cavernosa*). The names alone might tempt you to chew on them, just so you could rattle them off and impress your friends. That's what I did.

By late April, the first of the migratory caribou begin to abandon the boreal forest and trek to their traditional calving grounds in the arctic tundra. The pregnant cows are the first to leave, followed by the juveniles. The bulls, who often penetrate deepest into the boreal forest in winter, are the last to leave the tree line behind.

The caribou is the only boreal animal that commonly eats lichens, including this multi-branched Cladonia *species.*

Right: The damselfly is highly predatory and captures its insect prey on the wing.

Spring

WHEN I THINK OF SPRING IN THE BOREAL FOREST, I think of color and the crimson comb of a strutting spruce grouse or the cadmium yellow of a cluster of marsh marigolds. I think of the gilded eyes of a boreal toad as it peeps under starlight, and the smoky softness of a gray jay with its feathers fluffed up against the chill of dawn. From springs in the past, I remember the conversations of migrating geese as they wedge overhead, the temerity of a black bear testing the decaying ice of a frozen river, and the gray haze of an April storm as it races across the leaden surface of a lake. I also recall how the trunks of aspens bend in the might of a windstorm, creaking in protest, and how the familiar song of the olive-sided flycatcher and its urgent plea, "Quick, three beers," always makes me grin.

For many of the wild creatures of the taiga, spring is the season of renewal—a time of courtship and birth. It is also a time when goshawks hunt for distracted prey, when frozen frogs return to life, and when mother moose become ill-tempered guardians. These, and other stories, make spring a season of unexpected discoveries.

Above: The comb-like fringe on the bill of the northern shoveler enables it to strain tiny invertebrates and seeds from the water.

Croakers and Peepers

In 1758, Carolus Linnaeus, the father of animal classification, offered this scholarly description of frogs and toads. "Most amphibia are abhorrent because of their cold body, pale colour, cartilaginous skeleton, filthy skin, fierce aspect, calculating eye, offensive smell, harsh voice, squalid habitation and terrible venom." Linnaeus, like myself, studied and practised medicine and

The boreal toad relies on poisonous secretions from its skin glands for protection.

then, later in life, devoted his time to natural history. If he were alive today, I'm certain I could convince him that amphibians are not the foul and loathsome creatures he believed them to be, but rather a fascinating group of herptiles (amphibians and reptiles), with life strategies every bit as complex and intriguing as those of mammals or birds.

Life in the cold boreal forest is the ultimate challenge for any creature that depends upon the warmth of the sun to keep moving, as all amphibians do. For this very reason, the frogs and toads that inhabit the taiga are at the limit of their capabilities, and it is this that makes their study so fascinating.

Eight species of frogs and toads live in the taiga of North America: two chorus frogs, the spring peeper (*Pseudacris crucifer*) and striped chorus frog (*P. triseriata*); three true frogs, the wood frog (*Rana sylvatica*), the mink frog (*R. septentrionalis*), and the northern leopard frog (*R. pipiens*); and three toads, the American toad (*Bufo americanus*), the Canadian toad (*B. hemiophrys*), and the boreal toad (*B. boreas*). Each of these hardy herps uses one of three strategies to overwinter in the boreal forest: it either waterlogs, buries, or freezes itself.

The leopard frog and the mink frog overwinter at the bottom of lakes, ponds, and streams. In Lake Manitoba, ice fishermen have entangled hibernating leopard frogs in nets that were set in water 7 meters (23 ft.) deep and 18 kilometers (11 mi.) from shore. Where the lake bottoms are silty, the frogs may scoop out shallow pits in which to rest. Their metabolism slows down in winter, and the sluggish amphibians absorb enough oxygen through their skin and mouth lining to survive. Researcher Richard Cunjak went snorkeling in an Ontario streambed in winter, in water at a temperature of 1°C (34°F), to see how leopard frogs hibernated. The frogs wedged themselves between gravel and under rocks in the deepest part of the stream where the current was swift enough to flush away sediments that might smother them. Imagine the excitement and gossip a winter-snorkeling, frog-seeking scientist generated at the local coffee shop!

Toads are generally more terrestrial than frogs; for example, the three boreal toads go underground to escape from winter. Taiga toads

typically select pockets of sandy soil near their breeding ponds where they bury themselves, sometimes as deep as 1.2 meters (4 ft.). Overwintering toads may also use animal burrows or natural crevices that penetrate below the frost line. Where sandy soils are scarce, toads may hibernate communally. Near Fort Smith in the Northwest Territories, researcher Ernie Kuyt once found six hundred Canadian toads overwintering on a sandy hillside. Kuyt counted as many as forty-one toads in a square meter (thirty-four in a square yard), each in its own hole. Every year, hibernating toads in this northern taiga spent about eight and a half months underground.

The spring peeper, the wood frog, and the striped chorus frog adopt the most drastic tactic of all for overwintering; they freeze themselves alive. The trio spends winter on the soil surface, beneath the leaf litter and covered with a blanket of snow. Until the snow is deep enough to insulate the ground, the frogs can survive freezing temperatures as low as -8°C (18°F). If you uncovered one of these frozen frogs in midwinter, it would appear dead. Its eyes would be cloudy, it would not breathe or have a heartbeat, its limbs would be stiff, and its internal organs would be surrounded by a mass of ice. As much as two-thirds of the water in its body might be frozen solid. However, if you warmed the frog slowly, it would revive and eventually hop off the table.

Anyone who has ever had frostbite in a fingertip or toe has to wonder how the frogs survive such freezing. Tissue damage caused by freezing results when ice crystals form *inside* a cell, where they can puncture delicate membranes and generally upset the microscopic machinery of the cell, causing its death. It's a different story, however, when the ice crystals form *outside* cells, in the spaces between them, as they do in frogs. Here, the formation of ice causes no

The tiny striped chorus frog is often no larger than the end of a man's thumb.

damage. And this is how the frogs survive such freezing.

Amphibians in the taiga have a very narrow window of opportunity in which to mate, lay eggs, and grow tadpoles into tiny frogs and toadlets. The frost-free period in northern Manitoba, for example, is just seventy days. Because of this, boreal croakers and peepers get started with mating in early spring, usually in April, as soon as the ice begins to melt and sometimes when snow still covers the ground. The earliest species to breed are the three freeze-tolerant frogs. Like most taiga amphibians, the trio breed in marshy meadows, beaver ponds, and flooded ditches. The common feature for all these breeding sites is that they lack fish, which would prey heavily on vulnerable tadpoles. Even so, taiga tadpoles are still at risk for being eaten by predatory beetles, dragonfly larvae, and birds.

Male frogs and toads are the first to arrive at breeding pools in spring and upon arrival, immediately begin to call. The male of every species has its own unique voice. For example, the wood frog sounds like a quacking duck, and the tiny chorus frog reminds me of someone running a thumbnail along the teeth of a plastic comb. The American toad emits a musical trill lasting up to thirty seconds, and the leopard frog produces a snoring sound, often followed by a series of clucks, as if you had rubbed your fingertips over a rubber balloon. Advertising males may call hundreds or thousands of times in a single night. The thumbnail-sized spring peeper male peeps 4,500 times on any given night. Not surprisingly, calling is the most energetically expensive behavior in the life of a male frog or toad.

My most vivid memory of calling frogs involves a wolf. One April spring day in Saskatchewan's Prince Albert National Park, a solitary black wolf ran across the road, in front of my Jeep. The animal hesitated at the edge of the trees and stared back at me with its intense, amber eyes. An instant later, it dissolved into the shadows of the forest. In my mind, I could see the penetrating gaze of those golden eyes for many moments after the wolf had disappeared.

It was only after this vivid image began to fade that I first heard the quack and trill of hundreds of wood frogs and striped chorus frogs. It's peculiar how an intense experience with one of our senses can deaden the others. While the wolf was in view, I had been deaf to the din of the calling frogs.

Boreal amphibians are explosive breeders. Within days of the males' arrival, females converge on the breeding pool. Then the wrestling matches begin. Four or five male American toads may grapple for a single female and as many as ten male wood frogs may try to mate simultaneously with the same partner. Despite the mayhem, it's likely that the females do the choosing; they probably evaluate the quality of their partner before they let him mate. In toads and frogs, age may be a measure of superiority. In male toads, for example, older males are usually larger, and larger individuals have a deeper voice and are more attractive to females. Also, older male spring peepers call at a faster rate than their younger rivals do. As it turns out, female peepers prefer fast-talkers.

In one species of frog, the spring peeper, the intense competition for females has produced male cheaters. Cheaters never call and they never defend a territory. They simply loiter near a dominant male that is peeping and then try to highjack the first receptive female that comes by. Once a cheater latches onto a female, even a large dominant male has trouble dislodging him.

A successful male stays mounted on a female's back until she has laid all of her eggs, a process that may take several days. In toads, the breeding males have enlarged thumbs and roughened skin on their front feet to help them stay atop the female for the duration of these marathon mating sessions.

The three frogs that breed earliest in spring use a few tricks to help increase their egg survival rates. First, they usually lay their eggs in globular masses beneath the water surface. Thus, if a skim of ice forms overnight, none of the eggs is destroyed. Globular egg clusters are good at retaining heat absorbed from the sun and consequently remain warmer than the

This sedge marsh in northern Alberta has been used as a breeding site by wood frogs.

Red-sided garter snakes, frequent predators of boreal frogs, overwinter in crevices below the frost line.

surrounding water. As a result, the larvae develop faster. The wood frog uses a second tactic to increase the growth rate of its larvae: the females deposit their eggs in communal egg clumps. In fact, dozens of females may lay their eggs in the same area of a breeding pond. On a sunny day, the central egg masses may be 3.3 C° (6 F°) warmer than those on the edges, which enables some larvae to hatch even sooner.

Toads and frogs that lay their eggs later in the spring deposit them into water that is relatively warmer and therefore contains relatively less dissolved oxygen than earlier in the season. To ensure an adequate oxygen supply, they lay their eggs in layers or strings, which expose a larger surface area for oxygen absorption than the globular egg clusters described previously.

Breeding amphibians are conspicuous to predators, and many taiga animals feed on them, including black bears (*Ursus americanus*), red foxes (*Vulpes vulpes*), mink (*Mustela vison*), ravens (*Corvus corax*), gulls (*Larus* spp.), and owls (Order Strigiformes). Last spring, I watched a belted kingfisher (*Ceryle alcyon*) catch two boreal toads in less than thirty minutes.

In the taiga, the breeding season for amphibians usually ends in June, after which adults abandon the breeding pools as soon as they can, dispersing into forests, lakes, and ponds for safety.

Grouse—Strutting Their Stuff

In April 1999, I penciled these words into my field journal.

When my alarm rang at 5:00 this morning, I did a lot of moaning and groaning to encourage my weary bones out of bed. Today, I look and feel older than Elvis. Within an hour, I'm in the forest hiking towards "my" spruce grouse. This morning, the heavens can't decide whether to rain or snow, so they douse me with both. The patter of dripping water is the only sound in the gray gloom of dawn. As I approach the familiar forest opening, I hear the low-pitched drumming flight of the territorial male. He does one display flight after another. Up and down. Up and down. From the ground, he flies straight up to a branch, three to four meters overhead, turns around, and drops steeply downwards in a muffled whirr of wings. He selects another overhead branch and begins the sequence again. The energetic male makes a total of twenty flights in half as many minutes. The rival males on either side of him in the forest try to match his vigor, but neither can do it.

Next comes seven minutes of feeding on jack pine needles, then a few moments spent prospecting for grit in a patch of bare ground, and again, a dozen more drumming flights. After this second bout of flights, the grouse pauses to drink from a fallen poplar leaf in which a thimbleful of rainwater has collected. This doesn't slake his thirst, and he walks into the spruce bog nearby. I follow him closely and study his actions from just a few meters away. The grouse seems to have learned that the big noisy beast crashing through the underbrush behind him is not a threat and he ignores me. The thirsty bird stops at a large puddle of meltwater and begins to drink. I kneel on a soggy mound of sphagnum moss directly in front of him and slowly lean forward. The front of my camera lens is just half a meter from the grouse, and he continues to sip. I can't believe how tolerant he has become. It is a

Drumming male ruffed grouse run the risk of being detected and preyed upon by a northern goshawk.

When a female ruffed grouse approached the drumming log of this male, he immediately raised his ruff and began to strut.

precious moment of acceptance. This is the reason I photograph wildlife.

Grouse have fascinated me for years, and the boreal forest of North America is home to two of my favorite species, the spruce grouse (*Falcipennis canadensis*) and the ruffed grouse (*Bonasa umbellus*). The boreal forest of Eurasia has two equivalent species, the Siberian grouse (*Falcipennis falcipennis*) and the hazel grouse (*Bonasa bonasia*). As well, the taiga forests of Eurasia are vast enough to harbor four other kinds of grouse: the dapper Eurasian black grouse (*Tetrao tetrix*); the western and black-billed capercaillies (*Tetrao urogallus* and *T. parvirostris*), the two largest grouse in the

(*Populus* spp.) and birch (*Betula* spp.) trees, whereas a spruce grouse may feed almost exclusively on jack pine (*Pinus banksiana*) needles. Such low quality, fibrous diets are more typical of ruminants, such as moose and caribou, which employ a four-chambered stomach and process their food with the help of cellulose-digesting microbes. It turns out that grouse also ally themselves with cooperative microbes. Grouse have bacteria in their ceca, a pair of blind sacs attached at the junction between their small and large intestine. Before the start of winter, the bird's ceca lengthen by up to 40 percent. At the same time, its muscular gizzard doubles in size, to better pulverize its fibrous diet. In *Life in the Cold,* author Dr. Peter Marchand observes that grouse "have all the advantages of a ruminant without being limited to the ground ... and they have a virtually unlimited source of easily accessible food, requiring little energy to harvest."

It was a surprise to discover how little time grouse actually spend feeding in the winter. Typically, both taiga grouse in North America eat just twice a day, once at sunrise and again at dusk. At both times, they rely on the dim light to make them less conspicuous to predators. A spruce grouse takes about an hour to fill its large crop, but a ruffed grouse can do it in just fifteen to twenty-five minutes. Afterwards, both birds immediately disappear into snow cavities or dense conifers, where they conserve energy and digest their meal at a leisurely pace, hidden from the hungry gaze of goshawks (*Accipiter gentilis*) and owls (Order Strigiformes).

One of the great delights of spring in the taiga is the mating display of the male grouse. Each male struts his stuff in a different way. The most common and widespread grouse in the boreal forest is the ruffed grouse. This handsome woodland bird, with feathers that are a cryptic mix of browns, smokey grays, tans, and chestnut, is found wherever there are sizeable stands of aspen (*Populus tremuloides*) and balsam poplar (*Populus balsamifera*). Each year, between March and June, every self-respecting male ruffed grouse stakes out a drumming log and then rapidly beats his wings, over and over again,

world; and the willow ptarmigan (*Lagopus lagopus*). The ptarmigan also occurs in the taiga of North America, but primarily as a winter visitor from the Arctic.

Unlike most birds, grouse overwinter in the taiga with relative ease because of their ability to eat coarse plant food. During winter, a ruffed grouse feeds primarily on the buds of poplar to advertise his machismo to rival males and prospective female partners. In a typical drumming session, the energetic male beats his wings about fifty times, in a blur that lasts roughly eight to ten seconds (this is the kind of trivia that nature nerds thrive on). In late April, during the peak of the mating season, a testosterone-charged "ruffy" may drum like this every five to ten minutes for several hours at the beginning and end of each day, and sometimes during the night as well.

Every male ruffed grouse uses one main drumming site, commonly an old decayed log or exposed tree root, and one or two secondary sites. Usually the drumming sites are within 50 meters (55 yd.) of each other. Male ruffed grouse are true homebodies. Once they acquire a territory during their first or second winter, they may spend their entire lives within a 200-meter (219-yd.) radius of their main drumming log!

On a still morning, a human can hear the rhythmic drumming of a sex-starved drummer boy half a kilometer (0.3 mi.) away. Low frequency sounds penetrate vegetation better than do high frequency sounds, so the drumming of a ruffed grouse is well suited to broadcast the bird's message through the thick vegetation of the forest. Since the drumming is so clearly audible to the human ear, you would think it must also be easy for one of the grouse's mortal enemies, the great horned owl (*Bubo virginianus*), to locate and target the bird. However, it turns out that the frequency of the grouse's drumming is below the hearing threshold of the owl. Now that's smart planning.

Looking at a ruffed grouse melting into the brush or hopping off its drumming log, you might wonder where the bird got its common name. For the answer, you need the help of a female grouse. At the first glimpse of a prospective mate, the male fans his banded tail and raises the black ruffs on the sides of his neck. He hisses repeatedly as he struts and shakes his head. I always find it hard not to laugh, because the bird reminds me of an accelerating, miniature steam locomotive. The black feathers of his ruff have a greenish or purplish sheen, and when he shakes his head, the colors quiver beautifully.

The strutting dance of the spruce grouse is no less spectacular. The male droops his wings, cocks his tail, and raises the black feathers on his neck, framing his head and the engorged crimson combs above his eyes. As he struts forward, he alternately fans one side of his tail, then the other, giving the appearance of swaying. As his tail feathers rub against each other, they produce a quiet swishing sound. I watched the tame male spruce grouse described earlier strut like this for thirty minutes while he eyed a female perched 8 meters (26 ft.) overhead in a jack pine. I lay on the ground to be as unimposing as possible, and the cooperative male displayed so close to my face, I could ruffle his neck feathers with my breath. Wow!

The male spruce grouse performs his display flights during the same months that the ruffed grouse drums. Both birds concentrate their displays at dusk and dawn, sandwiching their activity and vulnerability between the nocturnal shift of predators and the diurnal ones. The noise of a display flight is a low-pitched whirr that lasts just a few seconds. The low frequency of the sound, like the drumming of a ruffed grouse, penetrates vegetation well and can be heard 100 meters (109 yd.) away, or more.

Spruce grouse prefer continuous tracts of coniferous forest, and each territorial male stakes out a small opening in the forest, where he struts and flutters. Often his display arena is just 15 meters (49 ft.) across. Such openings may be used by successive males for many years. I know of one display area, in northern Saskatchewan, that was used by a male in 1984 and again in 1989. I found another male there in 1997 and one in 1999 as well.

Male spruce grouse and male ruffed grouse are deadbeat dads. Once they contribute their seven to ten seconds of copulatory involvement, the hen is on her own to incubate the eggs and raise their family of chicks. The males continue their flamboyant displays well into early summer and mate with as many females as they can lure into their territory.

Early travelers in the boreal forest called the spruce grouse a "fool hen" because of its unwary nature.

One of the great delights of the spring in the taiga is the mating display of the male grouse. Each male struts his stuff in a different way.

Whiskey Jack Family Feuds

Nothing brightens a winter outing in the taiga like a pair of inquisitive gray jays (*Perisoreus canadensis*), whistling and gliding towards you through the snow-laden trees. These endearing robin-sized birds, with their soft smoky-gray plumage, trusting nature, and deep black eyes, have many common names: Canada jay, moose bird, camp robber, and carrion bird. My personal favorite is whiskey jack, an English distortion of the Algonquin Native name *whiskatjan*. The whiskey jack is a year-round resident of the North American boreal forest. Each mated pair defends a permanent territory and stays together for life. Since a jay may live for up to sixteen years, it will accept a new mate if its previous partner dies. The Siberian jay (*P. infaustus*) lives in the boreal forests of Eurasia and has the same personality and appeal as the whiskey jack. It is a snappy-looking, brownish-gray bird, with a rusty rump and flanks.

The whiskey jack is a member of the Corvid Family (Family Corvidae), which includes jays (*Cyanocitta* spp., *Perisoreus* spp.), nutcrackers (*Nucifraga* spp.), crows (*Corvus brachyrhyncos*), and ravens (*Corvus corax*). These intelligent, adaptable birds are the brainy ones of the avian world. None of them migrates in winter because each is creative enough in its feeding habits to ably survive the snow and cold. I've watched a gray jay delicately tear apart rose hips from the prickly rose (*Rosa acicularis*) in winter, eat the dried skin and pulp, and spit out the seeds. One of the jays' most novel foods is the blood-engorged winter tick (*Dermacentor albipictus*), which commonly infests moose (*Alces alces*) in the taiga. In one study in western Canada, the average number of winter ticks counted per moose was 32,500. Gray jays sometimes perch on a moose's back and peck around its anal area where winter ticks tend to concentrate. Aren't you glad you're not a hungry jay?

Like most corvids, the gray jay readily feeds on carrion. In winter, I've seen them scavenging the remains of grosbeaks (*Pinicola enucleator, Coccothraustes vespertinus*), crossbills (*Loxia leucoptera*), and red squirrels (*Tamiasciurus hudsonicus*) that were killed on the highway. I've also watched them pick over a skinned beaver (*Castor canadensis*) discarded by a trapper and peck the frozen carcass of a moose that faced off with a train. Probably my most interesting sighting was a pair of jays scavenging fat from the skin of a snowshoe hare (*Lepus americanus*) that had been ambushed by a lynx (*Lynx* spp.). Winter itself, however, is the biggest provider of carrion for jays. The lethal combination of persistent cold, deepening snow, and depleted energy reserves kills many boreal animals before spring can rescue them.

The gray jay is not only a successful scavenger, but also a capable predator of small mammals. In one winter study, small mammal remains were found in the stomachs of more than three-quarters of the jays examined between December and February. In Alberta, observers watched a whiskey jack kill two deer mice (*Peromyscus maniculatus*) on the surface of the snow, and in the Gaspé Peninsula of Quebec, a jay was seen killing a red-backed vole (*Clethrionomys* spp.) by pecking the animal's head repeatedly. It then flew off with the victim in its beak.

When food is scarce, the jay has one final option for feeding itself. It can recover food cached during the summer and autumn. This hoarding behavior is crucial to the bird's winter survival.

The whiskey jack is one of the earliest birds to nest in the boreal forest. Only the great horned owl (*Bubo virginianus*) may begin sooner. The jays start building nests in late February or early March, with most pairs having laid their eggs by the beginning of April. At this time of year, the eggs and naked hatchlings are at risk from the deadly cold, so the jays pay great attention to the construction of their nest and

This gray jay fluffs its feathers for insulation against the cold, -25°C weather.

Nothing brightens a winter outing in the taiga like a pair of inquisitive gray jays, whistling and gliding towards you through the snow-laden trees.

Ravens, which are related to jays, are among the most playful of birds; their aerial stunts commonly include swoops, rolls, and chases.

its location. First, they build a deep cup of tamarack (*Larix laricina*) and spruce (*Picea* spp.) twigs, shredded bark, and bits of old man's beard (*Usnea* spp.) and horsehair (*Bryoria* spp.) lichens, all woven together with spider and insect silk. The lining of the nest cup, which provides vital insulation, consists primarily of moose hair and feathers. Of two jay nests that were dissected, one contained 240 feathers and the other, 437, mostly those of grouse (*Falcipennis* spp., *Bonasa* spp). These birds build their nests close to the trunk of dense conifers for maximum protection from the weather and from predators such as martens (*Martes americana*), ravens, and red squirrels.

In March, temperatures can sometimes plummet to -29°C (-20°F), and no amount of nest insulation is enough to protect eggs exposed to such freezing cold. As a result, the female jay covers and warms her eggs almost continuously. For roughly three weeks, she hunkers down on the nest and leaves only a few times a day for stretches of four to eight minutes at a time. The male dutifully feeds his mate during this entire period.

In Quebec's La Vérendrye Game Reserve, young whiskey jacks are ready to leave the nest by April 11, when snow still covers the ground and 80 percent of the local migrant songbirds are still in the south. This rushed departure is the harbinger of a deadly family feud that erupts every June. Biologist Dan Strickland has been intrigued by gray jays for several decades, and his studies in Ontario's Algonquin Provincial Park have elucidated fascinating details about these birds and their feuds.

Most birds that live year-round in the taiga must contend with an uneven food supply—seasons of plenty, followed by months of meagre fare. They cope in one of three ways. The most common tactic is to escape, and more than three-quarters of the birds in the boreal forest migrate to warmer climes where food sources are more predictable. A small number of species stay and scrounge. Grouse, redpolls (*Carduelis* spp.), grosbeaks, crossbills, and woodpeckers (Family Picidae) fit into this category. The third group consists of hoarders that harvest the excess in times of abundance and store it for the lean months. Previously, I discussed how chickadees (*Parus* spp., *Poecile* spp.) do this; the same tactic is employed by most of the corvids, including the whiskey jack.

Adult gray jays begin hoarding in June and continue throughout summer and autumn. An individual may store hundreds of food items in a day, hiding each in a separate location, either in a cluster of conifer needles or jammed into a crack in some bark. In Alaska, the most common stored foods were insects, spiders, mushrooms, and berries. In fact, berries comprised 50 to 70 percent of the total, and the mushrooms even included poisonous varieties such as fly agaric (*Amanita muscaria*).

It's a little known fact that gray jays drool. Well, maybe they don't exactly drool, but they do produce lots of sticky saliva. Gray jays have exceptionally large salivary glands. Before they store a bolus of food, they shape it into a pellet by turning it over and over with their tongue and coating it with saliva. Dan Strickland speculates that the birds' saliva may contain anti-bacterial or other preservative qualities, although there have been no studies to verify this. He also wonders whether the volatile resins in conifer trees leach into the stored food, possibly increasing its longevity.

The foods that gray jays cache throughout the summer and autumn are the family's vital stores for the coming winter. Strickland believes this is why a family feud erupts in June, when the fledged chicks are roughly two months old and have begun to cache food on their own. Over a ten-day period, the clutch of three or four young battle with each other, and eventually one of them dominates its siblings, driving the others out of the family territory. In two-thirds of cases, the victor is a male offspring. Eighty percent of the vanquished young are dead by autumn. Of the few that survive, most are lucky females that locate a widowed adult male and become his new partner.

Why do the jays adopt such a seemingly wasteful reproductive strategy? Would they not produce more surviving offspring if the entire family stayed together until autumn, when the young have more experience, and then split up? Maybe so, but the winter survival of the adults might then be compromised. Strickland believes that a gray jay territory can normally support only two adults and a single offspring through the demands of winter. To do this, they must cache throughout the summer and autumn. If all young jays stayed with the family until autumn, they would learn the location of family caches and later steal them, possibly sentencing a parent to starvation. All animal behavior is shaped by natural selection, which perpetuates strategies that yield the maximum number of offspring. Family feuds and juvenile starvation are simply the best possible survival tactics for the whiskey jack.

Moose Mothers

The swivel-eared, droopy-nosed, dewlapped moose (*Alces alces*) is found throughout the taiga of North America and Eurasia. The moose is the largest member of the Deer Family (Family Cervidae), taller at the shoulder than a saddle horse. The average North American cow moose weighs around 350 kilograms (771 lb.), whereas the average bull tips the scales at roughly 430 kilograms (948 lb.). The largest moose live in Alaska and Yukon, where a big bull can weigh up to 700 kilograms (1,543 lb.). Among North American taiga animals, only the wood bison (*Bison bison athabascae*) is heavier than the moose.

The talented American wildlife artist Carl Rungius was considered by many to be the "Rembrandt of the moose." He jokingly claimed that his favorite subject to paint was "mooses in spruces." Indeed, in most people's minds, the moose is closely associated with the spruce (*Picea* spp.) forests of the taiga. However, moose numbers are surprisingly low in climax coniferous forests. Where the animal really thrives is in old taiga burns.

For much of the year, especially in winter, moose are browsers, feeding on the woody tips of willows (*Salix* spp.), paper birches (*Betula* spp.), and poplars (*Populus* spp.). In fact, the animal's common name is derived from the Algonkian Native name that means "bark stripper." Food abundance for moose reaches a peak ten to fifteen years after a boreal fire, and then gradually declines over the next twenty years, as the forest grows back. In Alaska, biologists measured the amount of edible browse available to moose three, ten, thirty, and ninety years after a forest fire. The totals were 37, 1399, 397, and 4 kilograms per hectare (33, 1246, 354, and 4 lb./acre), respectively. As the amount of available browse declined, so did moose numbers. Clearly, moose are dependent upon periodic taiga wildfires to create the habitat they need to survive.

The long-legged moose is the original all-terrain vehicle. With an average chest height of 105 centimeters (41 in.)—a third higher than both the caribou and the bison—moose can wade through deep snow more easily than any other hoofed animal in the taiga. They lift their legs from the snow vertically, rather than dragging them out at an angle, thus reducing the energy they burn when moving around. However, even in the best of habitats where food is plentiful, deep snow taxes the energy reserves of every moose; most lose between 15 and 20 percent of their body weight during a typical winter.

Russian researchers measured the winter ranges of moose under different snow conditions. During mild winters when the snow was shallow, moose used an area of 225 hectares (556 acres). When winters were moderate, they used slightly less than half this area, and during severe winters when the snow was deep, the haggard animals eked out an existence in just 5 hectares (12 acres). In many areas, moose react to deep snow by migrating to more favorable locations, sometimes moving as far as 200 kilometers (124 mi.), but more often less than 10 kilometers (6 mi.).

September and October is the breeding season for moose, with most giving birth in spring, in late May and early June. As in all mammals, the mother's nutritional condition greatly influences her ability to reproduce. When food is abundant and winter conditions are mild, as many as 90 percent of female moose will give birth to twins and on rare occasions, triplets. In contrast, when a pregnant moose has a poor diet and must endure a long winter with deep snow, she may give birth to only one calf or have no calf at all.

To avoid being detected by predators, moose typically select secluded locations for giving birth. Twice I found a mother moose with newborn calves, and both times the families were hidden in shrubby willow thickets. Islands are

another favorite birthing site. Veteran moose researcher Dr. Albert Franzmann reported on a pregnant female near his home in Alaska. The same female moose used a 1-hectare (2.5-acre) patch of spruce forest six years in a row. Each year, the cow would arrive at the calving site from two to seven days before she gave birth. Her young of the previous year were always with her, but the cow invariably prevented them from approaching closer than about 50 meters (164 ft.).

In 1998, Franzmann and his co-author Charles Schwartz wrote about the birthing process in their authoritative 733-page tome *The Ecology and Management of the North American Moose.*

Moose cows begin cleaning and drying their calves almost immediately after giving birth. The amniotic membranes and cotyledons [placenta] generally are consumed and the ground is cleaned of amniotic fluids. This may help eliminate scent that could otherwise attract predators or insects. Also, the licking of a calf almost certainly initiates essential bonding and it serves as a stimulus to activate within minutes the calf's struggle to rise and seek nourishment. Getting up and walking may take a calf several hours.

Members of the Deer Family rely on one of three strategies to protect their newborn young. The fawns of white-tailed deer (*Odocoileus virginianus*), mule deer (*O. hemionus*), and elk (*Cervus elaphus*) are "hiders." For the first few weeks of their lives, the young of these three species stay hidden by themselves, except when their mother returns to nurse them, a few times a day. The newborn hiders have a spotted coat to blend with the dappled light of their surroundings and emit very little scent. Whenever the young are nursing, the mother licks their groin.

This mother moose and her twin calves spent several hours resting in the meadow before they resumed browsing for willows.

A late winter storm may seriously damage the energy reserves of a pregnant moose and lead to a stillbirth.

To avoid being detected by predators, moose typically select secluded locations for giving birth. ... Islands are a favourite birthing site.

This stimulates them to void, and she then consumes their urine and feces. This maternal measure reduces dangerous telltale odors that could attract mammalian predators with their keen sense of smell.

Young caribou are "followers." Within an hour of birth, the young are strong enough to walk and follow their mother. Within days they can outrun a human and within weeks they can outrun a hungry wolf.

Newborn moose calves are neither hiders nor followers. Instead, they rely totally on their mothers to protect them. The mother's sharp hooves and aggressive disposition make her a formidable foe. For the first week after birth, a mother moose rarely strays farther than 50 meters (164 ft.) from her calves, and she may keep her offspring sequestered in the seclusion of the birthing area for up to three weeks.

The first six to eight weeks in a moose calf's life is the most vulnerable time, with predators posing the greatest threat. In Alaska, photographer Dave Fritts watched a grizzly (*Ursus arctos*) try to catch a young moose calf. The bear had killed the calf's twin a few days earlier. When the grizzly returned for the other calf, the mother charged it, lashing out with her front feet. The bear eventually limped away and left the mother and her calf in peace.

Another recorded moose-bear encounter involved a predatory black bear (*Ursus americanus*). When the bear first grabbed the moose calf, the calf began to bleat, and the mother moose, with her hackles raised and ears lowered, immediately charged the bear. The moose jumped on the bear's back with her heavy, sharp hooves and cut a deep gash across the animal's shoulder. The bear immediately freed the calf, which escaped.

Mother moose are not always so successful in protecting their young offspring. In different taiga areas, black bears and grizzlies may kill up to half the newborn moose calves; wolves (*Canis lupus*) may prey on an additional 18 percent. In some years, fewer than 17 percent of newborn moose calves survive this vulnerable period.

Many people have been charged and treed by an angry mother moose when they inadvertently wandered near her hidden calf. Because of this, I'm always especially wary of cow moose in the springtime. In late May 1985, while hiking in Prince Albert National Park in northern Saskatchewan, I suddenly found myself beside a moose, without warning. The leggy lady with lethal hooves and bloodshot eyes was below me on the hillside, barely 12 meters (40 ft.) away. Where was her calf? Had I come between the two of them, and was she about to charge? Within moments, the moose returned to stripping leaves with her fleshy lips, and I relaxed. I watched the mother for twenty minutes before her wobbly-legged calf finally slipped out from behind a curtain of alders. The newborn moose cast me a cautious glance before the two of them eased back into the tangle and out of sight.

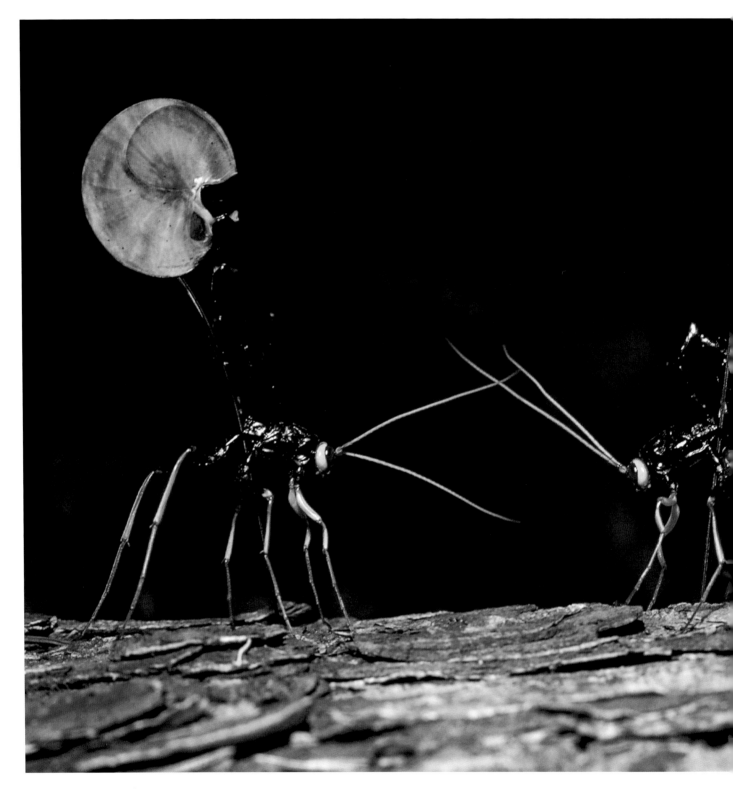

*Two female ichneumon wasps
drill the trunk of a dead pine,
searching for beetle larvae on
which to lay their eggs.*

Summer

WHEN I THINK OF SUMMER IN THE BOREAL FOREST, I think of water, warmth, and sunshine. I think of the thunder of waterfalls and rapids tumbling between the polished shoulders of rocky shorelines, and the clatter of lightning storms as they reaffirm the primal power of the heavens. From summers past, I remember the fearless scream of an eagle along countless deserted lakeshores, the constellation of dewdrops on a spider web at dawn, and the relaxing splash and drip of a paddle beside the gunwales of a canoe. I also recall the inviting warmth of a forest clearing carpeted with feathery mosses, the comforting whisper of aspen leaves conversing with the wind, and the tremor in my chest every time I photograph the elegance of a loon.

For many of the plants and animals of the taiga, summer is the season of recovery— a time to regain vitality and vigor. It is also a time when loons become lethal landowners, when the forest fills with the hum of the bloodthirsty, and when ospreys and eagles wage a tireless war. The stories of summer are deliciously complex and make it a season to relish.

Above: The stiff needles of the jack pine occur in clusters of two.

Right: When an insect lands on the blossom of the bog laurel, spring-loaded stamens are explosively released, thus dusting the insect with pollen.

Wail of the Wilderness

All loons, including this common loon, have red eyes, but the reason for this is still a mystery.

One of my most vivid memories of the common loon is associated with a canoe trip I made one summer in northern Quebec. It had rained for three days, my gear was soaked, and my spirits were soggier still. On the morning of the fourth day, I got up before sunrise and noticed the storm had passed. I took my coffee to the edge of the lake to savor the satisfying hush of morning and watch the day arrive. A thick, gray mist floated over the still surface of the water, and as the sun broke through the spruce trees, the mist was transformed into a swirling amber veil. A solitary loon surfaced near me, leaving a rippling wake of golden threads behind it. I wanted the bird to call, and when its plaintive wail swept over the lake, I knew it would be a perfect day.

No bird embodies the wild, hypnotic spirit of the taiga as much as the common loon (*Gavia immer*), whose haunting, quavering yodel echoes across lakes throughout the boreal forest of North America all summer long. Alaska and Yukon are the breeding grounds of a second, smaller species of North American loon, the Pacific loon (*G. pacifica*). In Eurasia, two additional species of loons, the arctic loon (*G. arctica*) and the red-throated loon (*G. stellata*), add their own bewitching wails and moans to the waters of the northern taiga.

In Europe, bird-watchers refer to loons as "divers," with good reason, for diving is one of the things these birds do best. Loons propel themselves underwater using their large webbed feet, unlike auks, which use their wings for this purpose. Because loons are foot-propelled divers, their legs are located as far to the rear of their bodies as possible. In this position, the

feet and legs give the birds greater power and maneuverability so that they can turn more quickly when chasing fish, and dive faster and deeper for their finned food. In 1952, author H. Wilcox published a detailed sixty-page scientific paper on the pelvic musculature of the common loon. Suffice it to say, Wilcox concluded that loons have strong legs.

The deepest recorded dive by a common loon is 81 meters (265 ft.), although most dives range from 3 to 4 meters (10 to 13 ft.) in depth. As well, diving loons have great underwater endurance. One arctic loon made 201 dives in three hours and twenty-five minutes, with only three pauses longer than a minute each. Most dives lasted from forty-five to fifty seconds.

Rear-positioned legs may be dandy for swimming underwater, but they are clumsy for walking on land. Only the small, red-throated loon can actually stand upright; other loons must slide on their chests and shove themselves along with their feet, much like a kid on a toboggan. The loon originally got its name from an old Norse word, *lomr* or *lom*, meaning lame, a reference to the awkward way these birds move on land. This lameness explains why all loons build their nests as close to the water's edge as possible—to keep their strolls on land to a minimum.

Each spring, common loons return to their boreal nesting lakes within a day or two of the ice melting. The anxious birds likely fly over their prospective territories repeatedly in the days beforehand, to assess the ice conditions. As soon as there is open water, they land and stake their claim on a territory. While they are waiting for the thaw, loons commonly congregate in the nearest open water, a wonderful spectacle to

This Pacific loon from northern Manitoba has just relieved its partner from sitting on their nest.

The red-throated loon is a common and widespread nesting species in the boreal forests of Eurasia.

Loon parents share in the month-long incubation, keeping the eggs covered for more than 90 percent of the time. The egg stage is the most vulnerable phase of the breeding cycle.

witness. On May 3, 1997, I counted twenty-eight common loons on a frozen lake in northern Saskatchewan. They shared a patch of open water not much larger than a football field. By May 7, all but four of the birds had left to occupy their nesting territories.

With loons, as with many migrants to the taiga, the early birds claim the best real estate. Small boreal lakes commonly hold a single pair of common loons, but some pairs may defend two or three small lakes in the same breeding season. In contrast, large taiga lakes with deeply indented shorelines and many islands may support over one hundred pairs of nesting loons.

Prime loon property has two qualities: good fishing and good nest sites. Eighty percent of a common loon's diet consists of fish, but loons also eat leeches (Class Hirudinea), caddis fly larvae (Order Trichoptera), dragonfly nymphs (Suborder Anisoptera), small crustaceans (Class Crustacea), and frogs (Order Anura). Disgruntled fishermen often claim that loons prey on important game fish, but the truth is, these birds most often target sluggish suckers (Family Catostomidae), sticklebacks (Family Gasterosteidae), sunfish (Family Centrarchidae), and perch (Family Percidae). During their four-month breeding season, an adult pair of loons and their two hungry chicks may gulp down 1,050 kilograms (2,315 lb.) of fins, flesh, and scales. Loons have a large muscular gizzard that grinds bones and other fish parts, aided by ten to twenty pea-sized stones that add grist to the muscle.

There are several records of loons dying when they tried to swallow a fish that was too large. In one case, a dead common loon was found with a 38-centimeter (15-in.) yellow perch (*Perca flavescens*) stuck in its throat. The fish's fins had jammed on the way down and were impossible to disengage. In northern Saskatchewan, I once watched a pair of loons struggle to swallow a 25-centimeter (10-in.) walleye (*Stizostedion vitreum*). I wrote in my field journal:

> For the last twenty minutes, the loons have repeatedly surfaced and dived with their

Loons often nest on remote, undisturbed lakes like this one in the Northwest Territories.

catch. I suggested to Aubrey that perhaps the birds were trying to drown their victim. The fish changed ownership a couple of times between the mates. Finally, one of them managed to get the fish's head fully inside its mouth, and after a dozen or so convulsive jerks, the fish disappeared down the bird's gullet. At one point, the loons left the dead fish floating on the water and a bald eagle swooped in to pilfer the catch. At the last moment, one of the loons saw the eagle coming, grabbed the fish, and with a squawk and a splash, escaped underwater with the prize.

The second important quality of good loon territory is a safe nesting site. Islands are ideal, especially those that offer good visibility, freedom from furry egg-snatchers such as black bears (*Ursus americanus*), wolves (*Canis lupus*), red foxes (*Vulpes vulpes*), and skunks (*Mephitis mephitis*), and protection from the wind and waves, which can flush an egg out of a loon's shallow nest. In some areas, 92 percent of loons build their nest on an island. Many pairs use the same island, year after year.

Competition for prime nesting territories may be intense; what's more, both sexes are fighters. An adult loon is a big powerful bird that can weigh over 5 kilograms (11 lb.) and measure nearly a meter (39 in.) from the end of its outstretched feet to the tip of its dagger-shaped bill. It's no surprise that battles between these well-armed birds are sometimes bloody, even deadly. A battling loon may try to drown an opponent by holding its head underwater or it may sneak underwater and impale an unwary intruder in the abdomen. If that fails, combatants may simply thrash it out on the water's surface, beating each other with their wings, bill-wrestling, and jabbing each other viciously with their beaks. Although fatalities are rare in these battles, more than half of the two hundred adult loons examined in a study in the northeastern United States had signs of old sternal punctures, wounds likely inflicted during such fights.

Loons do not limit their aggression to their own kind; they will attack anything they perceive as a competitor. There are reports of common loons attacking ducks (Family Anatidae), geese (Family Anatidae), and grebes (Family Podicipedidae), and even a beaver (*Castor canadensis*), a river otter (*Lontra canadensis*), and an irascible snapping turtle (*Chelydra serpentina*). In northern Minnesota, biologist Mark Sperry watched a territorial pair of loons kill, but not consume, two common goldeneye (*Bucephala clangula*) ducklings out of a brood of seven. The next year, in the same lake, the loons killed four out of six goldeneye ducklings, as well as a week-old ring-necked (*Aythya collaris*) duckling. The method of attack was always the same. Sperry wrote, "The loon approached underwater, surfaced next to the brood and grabbed one of the ducklings in its bill while the rest of the brood scattered."

One way that a pair of common loons may advertise its occupancy of a nesting territory is with guano, or as behavior biologists would say, with "stylized defecation." The territory owners climb up the shore, as far as 4 meters (13 ft.) from the water, aim their butt shoreward, and release a blast of whitewash. Sometimes a pair will defecate in sequence, one after the other; chicks also contribute routinely. The white guano is visible by humans from the air, so trespassing loons flying overhead can certainly see it as well. Veteran loon researcher Dr. Judith McIntyre thinks that the droppings may serve as a territory marker.

In June, female loons lay dark olive-brown eggs; three-quarters of females lay two eggs, whereas the rest lay a single egg. In her wonderfully detailed book *The Common Loon: Spirit of Northern Lakes*, McIntyre playfully quips,

"Anyone who has seen a loon egg is apt to remember it first for its size. Any female loon who has ever laid one no doubt remembers it for the same reason." Loon eggs average 150 grams (5.3 oz.) in weight. That's more than twice the weight of an extra-large hen's egg. Ouch!

Both parents share in the month-long incubation, keeping the eggs covered more than 90 percent of the time. The egg stage is the most vulnerable phase of the breeding cycle. Adults must shield and protect their clutch from the ravenous eyes of herring gulls (*Larus argentatus*), ravens (*Corvus corax*), and crows (*Corvus brachyrhyncos*). Downy chicks hatch within a day of each other, and within twenty-four hours, the youngsters leave the nest with their parents, never to return.

The aggressive nature of loons is apparent from the first day of a chick's life. The family lake may have a limited supply of small fish, aquatic insects, and other edible invertebrates. Siblings competing for these food items peck at each other vigorously during the first four days of life to establish a strict hierarchy. After that, the dominant chick is always fed first by its parents, and if food is scarce, the low-ranking chick may starve.

For the first two or three weeks of life, loon chicks are essentially freeloading floaters. They ride on their parents' backs and beg for food. Aboard a parent, the young loonlings are kept warm, shielded from underwater attacks by needle-toothed northern pike (*Esox lucius*), and safe from aerial strafing by hungry herring gulls and ravens.

If the young survive to this stage, they have about an 80 percent chance of fledging. At two months old, young loons catch half of their own food, and a month later, they are flying and totally self-sufficient. Juvenile loons migrate south after their parents have left the taiga for the winter, and it is likely that the family never reunites.

Bloodthirsty Bugs

In my wanderings through the taiga, I've squashed, smeared, splattered, and slapped legions of blackflies, yet I've never felt any great sense of satisfaction at exacting retribution. For author and avid fisherman Stewart Edward White, however, revenge was always sweet. In 1903 he enthused:

> [The blackfly] holds still to be killed. No frantic sweeps, no waving of arms, no muf-fled curses. You just place your finger calmly and firmly on the spot. You get him every time. In this is great, heart-lifting joy. It may be unholy joy, perhaps even vengeful, but it leaves the spirit ecstatic. The satisfaction of murdering the beast that has the nerve to light on you just as you are reeling in almost counterbalances the pain …

An ant drags a dead horsefly, many times its own weight, back to its underground nest.

In June and July, the taiga literally hums with thirsty insects searching for a hot-blooded donor. In North America, there are at least three hundred different species of bloodsucking insects in the boreal forest, but probably fewer than 10 percent of these regularly target humans. This fact has never been much of a comfort whenever I had to drop my drawers and squat in the woods during bug season.

Three main families of bloodthirsty insects stalk the taiga: blackflies (Family Simulidae), mosquitoes (Family Culicidae), and deerflies and horseflies (Family Tabanidae), also referred to as tabanids. Understanding the biology of these insects will never lessen the pain or itch of their bites, but there's always some merit in knowing the enemy and how he (or she, in this case) operates.

Blackfly eggs and larvae develop best in clear, cold, fast-running water that contains an abundance of oxygen. This perfectly describes the conditions in many rivers coursing through the taiga where these humpbacked little demons reproduce in numbers that challenge the imagination. In northern Quebec, for example, one rocky outcrop, no larger than an office desktop, was covered with sixteen million blackfly eggs. As soon as blackfly eggs hatch, the caterpillar-like larvae anchor themselves with silk to the river bottom and filter microorganisms from the currents flowing around them. After a couple of weeks and six to eight molts, the adults emerge from their silken cocoons and float to the surface on a bubble of gas. Once the little beasts break out of the water and dry their wings, they're ready to rock and roll.

The larvae of most tabanids and mosquitoes are also aquatic, but these insects prefer water conditions that are calm and quiet, sometimes even stagnant. Mosquito larvae commonly live in water-filled tree cavities, roadside puddles, and even inside discarded beer cans. All the larvae of these insects are filter feeders, similar to those of blackflies, seining the waters of their world for plankton with their brushy mouthparts.

The life cycle of one species of taiga mosquito, *Aedes vexans*, illustrates how wonderfully complex the lives of these insects can sometimes be. This little hummer will not hatch in just any old puddle of water; it needs water that is a soupy mix of organic debris. The organic debris guarantees a rich crop of microorganisms on which fast-growing mosquito larvae can feed.

But how does a dumb mosquito egg determine beforehand if there is enough debris in the water for it to hatch? Remarkably, the microscopic insect eggs can sense the level of dissolved oxygen in the water. When microorganisms graze on organic debris, they consume oxygen dissolved in the water. Consequently, low oxygen levels signal to the eggs that there are abundant microorganisms swarming in the water, the ideal feeding conditions for a newly hatched mosquito larva. My interest in natural history has always been fueled by such biological complexity; it is invariably a delight to discover another example of it.

As with blackflies, the larvae of mosquitoes and tabanids pass through a sequence of molts, and after a week or two, the adults emerge and fly off. The next stage in the life cycle of these bloodsucking insects is similar in all three groups. Typically, the males of each species swarm together at dusk or dawn, in closely packed clouds containing hundreds, and sometimes thousands, of individuals. They all face the same direction and hover to maintain their position. Although the swarms may gently drift up and down, they are relatively stationary, usually positioned over some visible marker point, such as a large boulder, a cluster of bushes, the crown of a tree, or a puddle of water. On the treeless prairies, I once watched a swarm of mosquitoes hover over a large, conspicuous cow pie. In *The Birder's Bug Book,* entomologist Dr. Gilbert Waldbauer describes how newly emerged virgin females enter the picture. "Females orient to the swarm, fly into it, and are almost immediately grabbed by a male, who takes his conquest out of the swarm and retreats with her to nearby vegetation, where they hide for the few moments that it takes him to inseminate her."

The males and females of most species of mosquitoes, blackflies, and tabanids drink plant nectar from flowering trees and shrubs to nourish themselves; only the females of these families search for a meal of blood. The nutrients contained in blood are used by the female to provision her eggs, and in many cases, the blood

meal enables her to lay a much larger clutch. For example, the females of one species of mosquito lay just twenty-five eggs when they feed solely on nectar, versus 150 eggs when they add blood to their diet.

Finding a victim is the first order of business for all bloodthirsty female insects. In the taiga, most follow the scent of carbon dioxide. The concentration of carbon dioxide in the exhaled breath of birds and mammals can be 150 times

greater than it is in the atmosphere, and insects have sensors in their antennae that can detect this gas from as much as 80 meters (87 yd.) downwind. Typically, the insects fly upwind in a zigzag course that holds them in the odor plume and systematically moves them closer to their quarry. Bloodsucking insects also rely on other odors to discriminate between potential hosts. The scent of lactic acid, acetone, and butanone, as well as urinary phenols, all products of

A nesting common loon is often plagued by black-flies, as the drop of blood on this bird's bill attests.

Rivers such as this one in Newfoundland are ideal breeding sites for black-flies.

Blackfly eggs and larvae develop best in clear, cold, fast-running water that contains an abundance of oxygen. This perfectly describes the conditions in many rivers coursing through the taiga.

metabolism, may contribute to the mix of odors that guides an insect to a meal. An extreme example of this is illustrated by the taiga blackfly *Simulium euryadminiculum,* which locates its favorite victim, the common loon (*Gavia* spp.), by targeting the odor of the bird's preen gland.

As a bloodsucking insect gets closer to its victim, it probably uses vision to guide its final approach. Researchers believe that some mos-

Bloodsucking insects feed in two different ways: mosquitoes puncture their victims, whereas blackflies and tabanids slash them. The mouthparts of the mosquito are modified into a tight cluster of long piercing blades that interconnect to form a tube through which the blood is sucked. In contrast, blackflies and tabanids crudely lacerate the victim's skin with their saw-edged mouthparts, then lap up the blood that oozes into the wound.

All blood-sucking insects feed rapidly to minimize the time when they are vulnerable to detection. Last summer I timed half a dozen mosquitoes as they fed on my arms and hands. All of them took between three and four minutes to bloat themselves with my blood. Blackflies may loiter longer and spend up to eight minutes lapping up a meal. Both of these biting insects can double or triple their weight when feeding on blood. Naturally, this seriously impairs their ability to fly and avoid predators such as flycatchers (Order Passeriformes), swallows (Family Hirundinidae), and dragonflies (Suborder Anisoptera). This is why most blood-bloated insects head for the protection of shrubbery or grass soon after they leave their victim. Blood is roughly 80 percent water, and within an hour or two, the insects excrete much of this, restoring their maneuverability.

One traveler to the taiga wrote, "Cold we could have endured, privation we were prepared for, but this horrible stew of flies ground upon our nerves till we were scarcely responsible for our actions." Countless accounts chronicle "the torture," "the torment," "the martyrdom," and "the madness" inflicted by bloodsucking insects on any and all who dared to enter the taiga during the summer bug season. Personally, I simply douse myself with bug dope and then try to execute as many of the little winged demons as I can. This, of course, is futile, and I always remember the old joke about the trapper who advises the pestered city slicker, "Listen son. Don't worry about the mosquito you just slapped on your arm. It's the thousand mourners who will show up for its funeral that you should worry about."

quitoes can see a silhouette up to 20 meters (66 ft.) away, especially if it contrasts strongly with the background. This is the reason light-colored clothing is often recommended for protection against these pests. In the moments before touchdown, the bloodthirsty she-devils may evaluate the body heat radiating from the soon-to-be-blood-donor and use the information to select a final landing site.

Ospreys and Eagles

In North America, two razor-taloned raptors share dominion over the water world of the taiga in summer: the osprey (*Pandion haliaetus*) and the bald eagle (*Haliaeetus leucocephalus*). Both are rapacious hunters of fish and mortal enemies. Where there is wilderness and isolation, the eagle rules, excluding the osprey from prime locations or forcing it to live where no eagle would. But when roads, power lines, dams, and development shatter the wild seclusion, the timid eagle retreats, and the fearless osprey inherits the vacated lakes and rivers. The life strategies of these splendid birds of prey are sometimes the same, sometimes not, but always a masterful match of creature to circumstance.

Among birds of prey, soaring consumes three to four times more energy than perching, and flapping flight burns ten to twelve times more. Thus, cost and benefit greatly determine the differing hunting strategies of the osprey and the eagle. The eagle is a frugal sit-and-wait hunter, perching beside the water and scanning the surface for a ripple, a fin, or the pallid flash of a fish corpse driven ashore by the wind. In northern Saskatchewan, eagles fish in this way for perch (Family Percidae), pike (Family Esocidae), whitefish (*Coregonus* spp.), walleye (*Stizostedion vitreum*), and fleshy-lipped suckers (*Moxostoma* spp.). All of these fish either swim close to the surface or spawn in shallow water, where an eagle can silently glide in and snatch one of them for dinner.

In contrast, the osprey is a spendthrift crash-and-splash hunter. Typically, it soars or flaps along the shoreline, 20 to 40 meters (66 to 131 ft.) above the water surface, searching for much the same prey as the eagle. When it spots a victim, the osprey's method of attack is unique among raptors: it plunge dives. Hovering on angled wings, the bird sights on its prey, then drops steeply like a feathered wedge. A heartbeat from the water, its legs swing forward, and it plunges through the surface, in a spray of liq-

uid highlights. For seconds, the bird vanishes while the empty water foams. Then it reappears, and with labored downstrokes, flaps itself free of the lake, clutching its silvery prize.

The osprey is well designed for such daredevil dives. Its long, featherless legs reach deeply into the water, to depths of a meter (3 ft.). To resist the penetrating rush of water, the bird's body plumage is dense, strongly barbed, and oily, and its nostrils are slitlike and closeable. Add to this strongly curved talons, a lightning grasp reflex of two-hundredths of a second, and spiny spiculed feet and you have a hunter that captures a meal in more dives than not.

Fishing on the wing increases the osprey's chances of locating prey, which offsets the extra energy it burns by hunting in such a strenuous manner. Ospreys are most successful at capturing prey when the weather is sunny and the water surface is calm. Visibility is optimal under these conditions, and the birds may succeed in up to 90 percent of their dives. Cloudy, windy conditions, however, can seriously impair an osprey's ability to locate fish, and during protracted periods of stormy weather, an osprey may lose weight, and nestlings may starve. When fishing conditions are poor, an osprey sometimes switches to other prey. In Labrador, for example, voles (*Microtus* spp.), meadow jumping mice (*Zapus hudsonius*), deer mice (*Peromyscus maniculatus*), and small songbirds comprised 12.5 percent of the local birds' diet.

The upper limit of what an osprey can eat in a single meal is about 300 grams (10.6 oz.), roughly a fifth of its body weight. Surplus food is discarded or carried about until the bird's crop empties enough to allow it to eat again. Researcher Dr. Alan Poole joked, "An osprey will sometimes clutch a fish carcass all day, feeding sporadically, reminding one of an old man unwilling to part with his cigar butt."

The bald eagle is three times heavier than the osprey, with an appetite to match its size.

A hungry eagle can stuff itself with a kilogram (2.2 lb.) of fish and bolt half of that in just four minutes. Such speed is not without purpose, as eagles are notorious thieves and will pirate a catch from each other any time they can. Juvenile eagles, two to four years of age, are the most notorious pirates, since they have yet to fully hone their fishing skills; for them, pilfering is often more successful than fishing.

Eagles also steal fish from loons (*Gavia* spp.), common mergansers (*Mergus mer-*

ganser), herring gulls (*Larus argentatus*), and even river otters (*Lontra canadensis*). The osprey is another habitual victim of these regal pirates. Benjamin Franklin, the celebrated statesman and inventor, deplored the eagle's thievery and lobbied strongly to prevent the bird from being chosen as the American national emblem. Franklin wrote:

He is a bird of bad moral character; he does not get his living honestly; you may see him

It takes four or five years for a bald eagle to acquire its full adult plumage.

In North America, two razor-taloned raptors share dominion over the water world of the taiga: the osprey and the bald eagle.

Most ospreys and bald eagles nest within half a kilometer of either a lake or river.

The white breast feathers of a perching osprey are visible from a great distance and advertise the bird's presence.

perched on some dead tree, where, too lazy to fish for himself, he watches the labor of the fishing-hawk [osprey]; when that diligent bird has at length taken a fish, and is bearing it to his nest for the support of his mate and young ones, the bald eagle pursues him and takes it from him.

Because eagles steal from ospreys so readily, it's no surprise that these two raptors are eternal enemies. Harassment is the osprey's only means of revenge, and screaming power dives, feather-rippling swoops, and persistence are its primary tactics. In one instance, an angry osprey swooped on an eagle seventeen times in five minutes, twice forcing the eagle to move to another perch, until the harried bird finally escaped into a tree where the osprey could not reach it. Neighboring ospreys will sometimes join forces to collectively mob a trespassing eagle and drive it away. Harassing an agitated eagle is a risky business. Eagles defend themselves by rolling over quickly in flight and trying to hook their tormentors with their feet. An eagle's foot can span 15 centimeters (6 in.), and the curved talon on its hind toe may be eight centimeters (3 in.) in length. Usually, an osprey's long narrow wings give it the vital maneuverability needed to avoid such lethal retaliation.

Many eagles and ospreys return to the frozen taiga in April, just as the lakes are beginning to thaw. When open water is scarce and the fishing poor, hungry eagles may scavenge the abandoned carcasses of wolf kills, feed on the drowned bodies of moose (*Alces alces*) or caribou (*Rangifer tarandus*) that fell through the ice in winter, or hunt weary waterfowl. Migrating ducks are especially vulnerable to predation at this time of the year because they are crowded together into small patches of open water.

Early one May, I watched an eagle target a victim in a large crowd of ducks. I was examining the eagle with binoculars when it dropped from its perch atop a towering white spruce (*Picea* spp.) and sailed towards a swimming flock of common goldeneyes (*Bucephala clangula*), mallards (*Anas platyrhynchos*), and common mergansers. On the first pass, the ducks scattered, squawking, splashing, and diving in every direction. The raptor banked sharply and swooped back over the water a second time. By now the eagle probably had its victim in sight—a female mallard that had dived underwater for safety. The hunter hovered about 6 meters (20 ft.) above the water and waited for the duck to resurface for a breath. It hovered there for almost fifteen seconds—a strenuous feat for a heavy bird with such large wings. Then the hapless mallard bobbed to the surface, and the eagle dropped on it instantly, sinking its lethal talons into the victim's back. For more than a minute, the exhausted eagle held the duck underwater and rested on the surface, with its wings outstretched.

Once the eagle had recovered, it labored to get aloft. The duck was still alive and flapped feebly, in a hopeless attempt to free itself. The powerful eagle flew to the shore about 40 meters (44 yd.) away and immediately began to feed on its struggling prey. With its massive hooked beak, the eagle pulled great clumps of feathers from the mallard's back, and after yanking a half dozen beakfuls, began to tear out bloody flesh. At this point, I could watch the killing no more. I've observed the raw edge of nature all my adult life, but sometimes the violent reality is more than I want to witness. Later, the eagle carried the duck's remains to his mate at their nest—crucial energy to fuel the next generation of eagles.

Muskeg—
The Beauty of the Bog

After an arduous field trip to Labrador, artist John James Audubon called the muskeg areas of the taiga "a poor rugged miserable country." Writer Barbara Moon described muskeg as "a treacherous, sucking, ill-smelling bog of peaty muck, spongy sphagnum moss and standing water." Author Tom Alderman carried the denigration further. In 1965 he opined, "It just lies there, smeared across Canada like a leprosy. … In summer, it's a rotting mushland of blackflies and mosquitoes, and the odor is akin to backed-up septic tanks. In winter, it's an eerie half-world, a frozen lifeless wasteland defying civilization." Alderman was writing for the *Imperial Oil Review* and his slanderous eye was clearly focused on the sacred cow of corporate profit. "Beneath this matter could be a wealth of crude oil, natural gas and minerals. Even more fascinating, locked within its slime is a future to be made from chemicals and synthetic materials." He concluded, "Muskeg is both a prize and a plague, a crazy mixed up quagmire with a split personality."

Polishing the dismal image of muskeg may be an impossible task, but I love the challenge. In summer, muskeg *is* indeed soggy and insect-ridden. For the novice, it *can be* a terrifying maze of tamarack (*Larix laricina*) and spindly spruce (*Picea* spp.), nothing more than an inhospitable wetland. But beyond all of this is a landscape of inspiring beauty and rich biological wealth. Here you can discover puzzling carnivorous plants, delicate showy orchids (Family Orchidaceae), and secretive nesting northern hawk owls (*Surnia ulula*). Above the hum of insects, you can hear the bugling duets of courting sandhill cranes (*Grus canadensis*), the primal howl of wolves (*Canis lupus*), and the distant siren call of lovesick loons (*Gavia* spp.). Muskeg is not for the meek or timid, but the challenges it poses are filled with reward.

Muskeg is a colloquial term for a wetland where peat—the undecayed remnants of plants—accumulate on the ground. In most environments, dead plants are continually broken down by bacteria, fungi, and fire, and the nutrients recycled, with the annual cycle of decay balancing the yearly production of new growth. In muskeg, the cold wet conditions slow the normal process of decay or halt it completely, and peat accumulates as a result. By some estimates, as much as 24,650 kilograms per hectare (22,000 lb./acre) of new peat may be added every year to some areas of muskeg.

In North America, the distribution of muskeg more or less follows that of the boreal forest. Canada has over a million square kilometers (386,100 sq. mi.) of the soggy stuff, most of it lying over the bedrock of the Canadian Shield. The densest tract of muskeg stretches as a broad belt up to 500 kilometers (310 mi.) wide, extending from the southern tip of James Bay northwest for 2,900 kilometers (1,800 mi.) or more, across northern Manitoba, into the Northwest Territories. It ends around Great Bear Lake. Much of Newfoundland is also comprised of dense muskeg. Elsewhere in the taiga, the distribution of muskeg is patchy, usually occurring in scattered blocks wherever permafrost and (or) the local topography prevents or slows drainage, leaving the ground soggy or flooded.

There are two types of muskeg: fens and bogs. In fens, the flow of groundwater is sluggish, but sufficient for replenishing important nutrients such as calcium and magnesium. Sedges (Family Cyperaceae) are usually the dominant plants in fens, although heath shrubs (Family Ericaceae) and willows (*Salix* spp.) often grow in the slightly drier areas, with solitary black spruces (*Picea mariana*) and tamaracks (*Larix laricina*) scattered around the edges. Whereas fens are the mineral-rich half of the muskeg world, bogs are the impoverished ones. In bogs, mineral levels are extremely low, and the groundwater is stagnant, with little internal seepage to replenish nutrients. Sphagnum mosses (*Sphagnum* spp.) are frequently the dominant plants in bogs, but as in fens, tenacious heath shrubs and black spruce also occur in the less waterlogged areas. Sphagnum mosses are better known as peat moss, the common soil amendment used by gardeners. According to *The Field Guide to the Peat Mosses of Boreal North America* by Cyrus McQueen, there are forty-eight species of sphagnum in the taiga of the continent, and half of them may grow in just one small bog. It cost me over $40.00 to learn this fact. I bought this guidebook reasoning any nature nerd worth his salt should know as many muskeg minutiae as possible.

Sphagnum mosses have the unique ability to absorb moisture better than a sponge and to hold up to twenty times their weight in water. This quality is what makes peat moss so popular with gardeners. But the absorbency of these mosses is useful in other ways. When I practised medicine in northern Ontario, some of my patients were Ojibwa and Cree trappers who lived in the bush for much of the year. From them I learned some traditional Native uses of sphagnum. The people collected the moss in the summer, and once it was dried, it was used as biodegradable menstrual pads and diapers. I never once saw a Native baby whose bottom was swaddled in sphagnum develop a diaper rash. In the past, dried sphagnum was also used as a dressing for wounds. The mosses contain phenols, which have antiseptic properties. I once treated a Native man who had seriously slashed his leg with an axe while he was out hunting on the land. He packed the wound with fresh sphagnum and continued to hunt for another week. When I saw him later, the wound was clean and healing well, with no signs of infection.

Wildlife may inadvertently capitalize on the antiseptic properties of sphagnum. In a study in the Norwegian taiga, half of the animal carcasses cached by brown bears (*Ursus arctos*) were covered with sphagnum moss. Although the researchers could not prove that the bears preferred sphagnum as a covering material, the mosses may nonetheless have worked as a preservative.

Four to 5 centimeters (1.5 to 2 in.) below the cold wet surface of a patch of peat, there is no oxygen to be found. Decomposition is slow, and nutrients are scarce. Without oxygen, the peat

The showy lady's slipper is the largest orchid in the taiga of North America.

decays along anaerobic pathways that produce skatole, indole, hydrogen sulfide, and mercaptans—the familiar odors of putrefaction. When you slog through a bog, you stir up the peat and release this distinctive "swamp gas." The moment you start to enjoy the smell, you know it's time to head home.

Three of my favorite groups of boreal plants grow in the difficult conditions imposed by muskeg. At the top of the list are the meat-eating plants that compensate for the scarcity of available nutrients by luring, capturing, and then consuming insects and other tiny invertebrates. I devote the next essay in this book entirely to these carnivorous plants and all the morbid details of their lives (see page 108). Orchids, a more genteel group of plants, bring style and beauty to the bog. Worldwide, there are perhaps thirty thousand different species of wild orchids, the greatest number of which grow in the tropics. Even so, several dozen species enliven the muskeg with their delicate floral designs, patterned petals, and subtle fragrance. None of these exquisite plants is very numerous, and all are rewarding to find.

The common name orchid comes from the Greek word *orchis*, which means testicle. In the minds of the ancients who named these plants, the swollen tubers in orchid roots resemble testicles, and some of the blossoms, especially those of lady's slippers, look like a scrotum. In those days, people believed in the Doctrine of Signatures, which explained that when a plant resembles a part of the human anatomy, it is a signal of its power and medicinal value for that body part. As a result of this belief, orchids have a long history as erotic charms and aphrodisiacs. Perhaps our modern use of orchids in courtship and marriage hearkens back to those ancient beliefs?

The heath plants, or ericads, are the most conspicuous of my trio of muskeg favorites. They include some familiar plants, such as labrador tea (*Ledum groenlandicum*), blueberries (*Vaccinium* spp.), and cranberries (*Vaccinium* spp.), and some less familiar (though sometimes more common) species,

Sphagnum moss is one of the dominant plants that grow in the waterlogged soils of the bog.

The sole breeding ground of the endangered whooping crane is a remote area of muskeg in Wood Buffalo National Park.

such as leatherleaf (*Chamaedaphne calyculata*), bog rosemary (*Andromeda glaucophylla*), and the beautiful crimson-flowered bog laurel (*Kalmia polifolia*).

The ericads' tenacity for growth, and the visible ways they have adapted to the extreme conditions of the muskeg are what interest me most about these plants. The success of the ericads relies on two main adaptations: their capacity to conserve nutrients and their resistance to moisture loss. A number of the heath shrubs retain their leaves for several growing seasons, sometimes for four years or more. During the winter, their leaves may turn brown and look dead, but as soon as the warm weather returns, they flush with green and quickly begin to photosynthesize. By retaining their leaves for several summers, the ericads reduce the volume of nutrients they must extract each year from the cold, soggy, impoverished peat of the muskeg to produce new leaves. You will recall that conifers retain their needles through the winter for the same energy-conserving reasons.

Typically, muskeg is waterlogged, so at first it's quite surprising to learn that the leaves of many heath plants have adapted to conserve moisture. For example, ericad leaves have a thick waxy coating, curled edges in many species, and on the undersides of some, a thick fuzzy covering called *tomentum*. The wax retards evaporation, and the curled leaf edges and tomentum dampen the movement of air around leaf pores, potential sites for moisture loss. Such adaptations, which are typical of desert plants, are equally important for muskeg plants.

Even though muskeg is usually saturated with water, for much of the year that water is frozen and unavailable to the roots of plants. In May out on the muskeg, air temperatures may climb into the mid-20°C (mid-70°F) range, yet ericad roots may still be locked in ice. It's likely the plants would shrivel and die from desiccation if their leaves were not adapted to prevent excessive evaporation.

One September, I flew in a small plane over a patch of muskeg in northern Ontario. From the

air, the bog was an inviting panoply of golden tamaracks and willows, crimson and purple sphagnum, and shimmering blue water. Later the same day, I hiked into the bog to savor it more closely. For several hours, I sat and listened to the wind conversing with the land, soaking up the beauty and secrets the muskeg had to offer. The English poet William Wordsworth may never have hiked on muskeg but he certainly knew how I felt that day when he wrote, "Nature never betrays the heart that loves her sincerely."

Small bees and flies pollinate the tiny, delicate blossoms of the round-leaved orchid.

Meat-eating Plants

Throughout life, my interest in the natural world has always focused on creatures with fur, feathers, or scales. Whenever the topic of botany came up, I invariably pleaded leniency for my disinterest and promised to concentrate on the study of plants in the seventh decade of my life. Until then, my heart belonged to critters. All that changed when I discovered sundews, pitcher plants, butterworts, and bladderworts. These are carnivorous plants, plants with an appetite for blood—definitely my kind of plants. As it happens, the summer muskeg of the boreal forest is a great place to study them.

All plants require nitrogen for growth. In bogs and fens, much of the nitrogen is locked inside undecayed peat and is unavailable for growing plants. By capturing and digesting animal prey, meat-eating plants increase their intake of nitrogen, which allows them to produce a larger crop of seeds. Each of the four groups of carnivorous plants adopts one of three strategies to capture its victims: suction trap, flypaper trap, or pitfall trap.

The bladderworts (*Utricularia* spp.) employ a suction trap. Roughly half a dozen species of these aquatic plants grow in the muskeg of North America, and several of the same species also occur in Eurasia. All of these plants have leaves that are modified into tiny bladders, 0.3 to 0.5 centimeters (0.1 to 0.2 in.) long. The plant transforms these underwater bladders into deadly trapping devices by actively pumping 90 percent of the water out of them. The walls of the bladder are then under tension to re-expand. Alluring secretions from the lip of the bladder tempt microscopic animals such as water fleas (*Daphnia* spp.), copepods (*Cyclops* spp.), scuds (Order Amphipoda), roundworms (Phylum Nematoda), and protozoans (Phylum Protozoa) to approach. When one of these tiny creatures bumps a trigger hair, the bladder door snaps open, and the animal is sucked inside with the rush of water. The door closes immediately,

trapping the victim inside. The bladder then secretes acid and digestive enzymes, which combine with bacteria to decompose the prey. Afterwards, the dissolved nutrients are absorbed. In waters rich with invertebrates, a single bladderwort plant may trap exceptional numbers of animals. For example, a large common bladderwort (*Utricularia vulgaris*) may be armed with six hundred underwater bladders, which collectively capture one hundred and fifty thousand victims in a summer!

The sundews (*Drosera* spp.) and the butterworts (*Pinguicula* spp.) use sticky flypaper traps to secure food. In both of these carnivorous groups, the small leaves of the plant are arranged in rosettes, close to the ground. The upper surfaces of the leaves are covered with glands that secrete a sticky mucilage, one of the most powerful adhesives in nature. A single leaf may have 250 of these glands. Neither group produces sugary nectar to lure potential victims closer, but even so, tiny flies are possibly fooled by the glistening appearance of the adhesive and land to investigate. Butterworts have a faint fungal smell. This odor, and others undetectable to the human nose, may also tempt and attract victims. When prey eventually lands on a leaf surface, the sticky tentacles fold over the creature, enveloping it and flooding it with digestive enzymes. At the same time, the edges of the leaf may curl inwards, further trapping the insect and preventing the loss of nutritive fluid. In sundews, this process may happen in a matter of minutes, but in butterworts, it may take hours. It is critical that digestion and absorption occur rapidly, before rainwater washes away the precious nutrients. Typical victims include mosquitoes (Family Culicidae), flying aphids (Family Aphididae), midges (Family Chironomidae), and springtails (Order Collembola), but nothing as big as a house fly (Family Muscidae). As many as five hundred tiny flies have been counted on the leaves of a single butterwort plant. All that

The pitcher plant is widespread in sphagnum bogs from Newfoundland to northeastern British Columbia.

The glistening mucilage on the leaf tentacles of the round-leaved sundew traps unsuspecting insects and digests their bodies.

remain of the victims afterwards are empty exoskeletons that blow away in the wind.

Charles Darwin was fascinated by meat-eating plants and in 1875, he published his book *Insectivorous Plants*. Darwin experimented with sundews and tried to trick the plants into reacting to a tiny pebble dropped on a leaf. He discovered that the plants would only respond to organic materials, such as human hair or a piece of meat.

The pitfall trap employed by the pitcher plant (*Sarracenia purpurea*) is my favorite carnivorous plant strategy. In all of the meat-eating plants, it's the leaves that are adapted for carnivory. The flask-shaped leaves of the pitcher plant, 10 to 30 centimeters (4 to 12 in.) in length, fill with rainwater and fluids secreted by the leaf. Colorful crimson and purple veins on the hood of the pitcher serve as nectar guides, for attracting a wide variety of prey and encouraging it to land. In a Michigan study, researchers examined the prey in 214 pitchers and found forty-nine different families of insects represented, most of them species of flies. The hood of the pitcher is covered with stiff, inward-pointing hairs that direct the victim towards the mouth of the pitcher, where nectar glands entice it deeper inside. The upper walls of the pitcher are covered with a slippery wax that prevents prey from getting a foothold and backtracking, and any victim lured this far usually tumbles into the pool of liquid at the bottom. The narrow neck of the pitcher plant probably prevents insects from flying to freedom. Captives eventually become exhausted and drown. Next, glands in the walls of the pitcher plant release caustic hydrogen ions, which make the liquid pool as acidic as vinegar. The process of digestion and absorption of prey then begins. If this was the end of the story, it would be remarkable enough, but in fact, this is just the beginning of an incredible tale of murder, cannibalism, and scavenging. Ecologist Dr. Stephen Heard went to Newfoundland, where pitcher plants are abundant, to unravel the amazing story that follows.

Hidden within tiny pools inside the leaves of the pitcher plant is an entire ecosystem dominated by creatures that live nowhere else on Earth. Researcher Heard identified five different invertebrates inhabiting the pitcher plants in Newfoundland. A rotifer (*Habrotracha rosa*) is the smallest and most numerous creature. One pitcher might contain a thousand of these microscopic filter feeders, which spread from pitcher to pitcher by adhering to the legs and bodies of passing midges and mosquitoes. Another pitcher inhabitant is a predatory mite (*Sarraceniopus gibsoni*), which feeds on bacteria and particulate matter filtered from the soupy liquid. Between thirty and two hundred mites were present in most pitcher plants that Heard examined. Like the rotifer, the mite completes its entire life cycle inside the pitcher and then attaches itself to passing insects to disperse.

The largest of the pitcher plant critters is the maggot of a flesh fly (*Blaesoxipha fletcheri*), which is 1 centimeter (0.5 in.) long. In late summer, the female flies give birth to live larvae inside the pitchers. The newborn larvae dangle from the underside of the water surface and feed on the floating carcasses of drowned and dying insects. If more than one of these voracious and highly aggressive maggots is present, they wrestle and fight with each other, until only one survives. The victor then cannibalizes the drowned loser. In autumn, the maggots crawl out of the pitchers to complete their life cycle; they overwinter in the sphagnum moss beneath the plants.

Another insect Dr. Heard found inside the pitcher plant was the midge *Metriocnemus knabi*. He counted from ten to fifty larvae of these mosquito look-alikes in the bottom sludge of almost every pitcher. The larvae, which are 0.6 centimeters (0.25 in.) long, use their sharp jaws to tear apart dead insect material that floats down from the surface film. The grubs overwinter frozen in the ice of the pitcher, emerging as adults in late summer. After this, they quickly mate and find new pitchers in which to lay their eggs.

In the insect-ridden world of the muskeg, it's fitting that one of the creatures that live in pitcher plants, the mosquito *Wyeomyia smithii*, is unique to the plant. The larvae of the pitcher mosquito cling to the surface film of the water and filter bacteria and particulate matter. These larvae, like those of the midge, overwinter frozen inside the pitchers, emerging as adults in midsummer. The adult female mosquitoes feed on nectar, or not at all, and they never feed on blood.

Dr. Heard describes the community of invertebrates living inside a pitcher plant as a "processing chain." He writes in summary:

> Each species is ultimately dependent upon the same resource; that is, upon the prey caught by the pitchers. However, each uses it in a different state of physical decomposition, and each participates in decomposing and processing the resource. Floating carcasses are used by the sarcophagid [flesh fly], sunken biomass by the midge, suspended particulates by the mosquito, and dissolved nutrients by the plant.

Hell-divers and Water Witches

The horned grebe dives for its meals and hunts small fish, tadpoles, leeches, and other aquatic invertebrates.

In 1997, the highly respected International Union for Conservation of Nature and Natural Resources (IUCN), in Switzerland, released a status report on the grebes of the world. The report began, "Grebes have been around for 40 million years. When you look at a grebe you are looking at something whose genetic identity is ten to twenty times longer than your own. This is a good way to start thinking about grebes and their place on the planet we fondly imagine as ours." In the same report, author Dr. N. J. Collar enthused:

To witness their courtship—those wonderful water-dances when two birds tread water face-to-face waving water-weed in their bills, or rush side-by-side across the water, bodies upright and calling wildly—is less to learn anything of *their* lives than to sense the crippling limitation of *ours*.

Grebes are specialized waterbirds that inhabit marshes and shallow lakes throughout much of the boreal forest worldwide. Although they resemble ducks in size, grebes have a slen-

der pointed bill, not a flat one as most ducks have. All grebes have relatively small wings and weak pectoral muscles for their body size, so they must patter across the surface of the water for quite some distance before they gain enough speed to become airborne. During the summer nesting season, grebes rarely fly. They prefer to dive or swim to escape from danger, or simply to disappear into thick stands of cattails (*Typhus* spp.) or bulrushes (*Scirpus* spp.). In fact, some species may be almost flightless during the summer, as their breast muscles atrophy to lessen the energy cost of maintaining something they rarely use. These skilled diving birds have the uncanny ability to disappear underwater without a ripple, and the speed with which they do this has earned them several colorful nicknames: devil-diver, hell-diver, and water witch.

In summer, two species of grebes are common to the boreal regions of both North America and Eurasia: the red-necked grebe (*Podiceps grisegena*) and the horned grebe (*P. auritus*). A third species of grebe, the great crested grebe (*P. cristatus*), occurs in the boreal forests of western Eurasia, whereas the diminutive pied-billed grebe (*Podilymbus podiceps*) is found in the taiga of North America.

The legs on a grebe are positioned at the extreme rear of its body, similar to loons' legs. Although this is a great advantage for diving, it makes walking on land extremely awkward. For this reason, grebes build floating nests of soggy water weeds, which they anchor to nearby cattails or bulrushes. Typically, grebes lay three to six chalky-white eggs, with both parents sharing incubation. Anytime an adult leaves the nest, it covers the eggs with rotting nesting material to keep them from overheating and to hide them from the sharp eyes of gulls (*Larus* spp.) and ravens (*Corvus corax*). As a result, the pale eggs are soon stained brown.

The embryos in most birds, including grebes, become fixed to the inside of their eggshells roughly half way through the incubation period. From then on, an embryo may be incubated right side up, sideways, or upside down. For grebes, this is not a problem until hatching time arrives. A sodden grebe nest may have a shallow puddle of water in the bottom of it, and if a chick is upside down when it hatches, it might drown before it can completely break out of the eggshell. The solution? Talk to the parents. Twenty-four to forty-eight hours before hatching, grebe chicks begin to peep inside the shell; the rate at which they do this varies. This behavior has been studied in the eared grebe (*Podiceps nigricollis*), a close relative of the boreal grebes, and it's quite likely that it occurs in the taiga species as well. When researchers turned eared grebe eggs so that the chicks were upside down and in a dangerous position, the chicks peeped more rapidly; they only stopped once they were returned to an upright position. Normally, grebe parents rotate their eggs at least once a day. The chick vocalizations in the final

Red-necked grebes are loudmouthed lovers; mated pairs sing a duet that sounds like a noisy horse whinny.

days before hatching may signal parents when they are in the proper position, and thus, when to stop turning their eggs.

Young grebe chicks leave the nest within a day of hatching and immediately climb onto a parent's back. Here, snuggled under a wing, they

A red-necked grebe returns to its nest and removes the soggy vegetation covering its clutch of five eggs.

Grebes are specialized waterbirds that inhabit marshes and shallow lakes throughout much of the boreal forest.

find shelter from cold spring downpours and protection from the heat of the sun. For the first week of their life, hatchlings eat, sleep, and travel almost continuously on a parent's back, even when the adult dives underwater to search for aquatic beetles, dragonfly nymphs, tadpoles, leeches, and small fish. Periodically, the chicks are shaken off into the water, where they can defecate.

Young grebes, the original skinheads, have a novel way of communicating with their parents. In all boreal species except the pied-billed grebe, the boldly striped youngsters have a triangular patch of bare skin on the top of their head. The featherless skin flushes from fleshy pink or pale yellow to bright crimson whenever the youngsters get excited, especially when begging for food. Biologists suspect that the parents use this flush of color to monitor a chick's hunger.

A grebe's waterproof plumage is especially soft and thick; in fact, a grebe may have over twenty thousand feathers on its body. In the late 1800s, grebes were heavily hunted. Their thick feathers were marketed as "grebe fur" and used to adorn the capes and hats of stylish women of the era. Long before the grebe furriers came along, the birds were using their own feathers in a way that is unique among birds. Grebes pluck and eat their feathers on a regular basis. In fact, over half the contents of a horned grebe's stomach may consist of feathers. Researchers speculate that the ingested feathers may protect a grebe's stomach from sharp fish bones. Another possibility is that the feathers act as a filter, preventing sharp food debris from entering the intestine. Eventually, a wad of feathers, bones, and debris is regurgitated and spit out as a harmless pellet. No matter why grebes feast on feathers, this behavior is crucial to them. Sometimes grebe parents feed their day-old chicks small feathers, even before they give them any food!

*The powerful talons and fierce
disposition of the great horned
owl make it one of the most
deadly taiga raptors.*

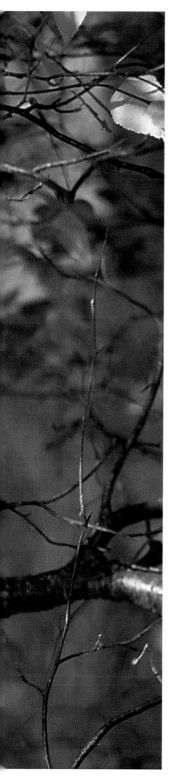

Autumn

WHEN I THINK OF AUTUMN IN THE BOREAL FOREST, I think of the rich damp smell of fallen leaves, the howl of coyotes and wolves as pups learn to wander, and the gurgle of beaver dams strengthened for the months ahead. I think of fanciful mushrooms with colorful caps and gills of ivory, and the fallen needles of tamaracks littering the ground like slivers of gold. From autumns past, I remember how moose noiselessly strip the yellowing leaves from a willow branch, and how a porcupine nibbles on the last rose hips of the season. I also recall how tireless a family of beavers can be as they cache branches for winter, how tame a boreal owl is when hidden in the thickness of a spruce tree, and how silent the lakes seem after the loons and waterfowl leave.

For most of the wild creatures of the taiga, autumn is the season of escape or preparation. Many will flee, but just as many will remain behind. It is also a time for bears to dig their dens, for porcupines to battle, and for squirrels to hoard magical mushrooms. The stories of autumn are some of the most surprising of any season in the taiga.

Above: A grove of aspens may comprise a single plant or clone; all members of the clone will turn yellow at the same time.

Right: Moose often eat water lilies and other aquatic plants, savoring the saltiness of the vegetation.

Flowers on the Wing

Many bird-watchers dream of a spring pilgrimage to Point Pelee National Park, in the same way that devout Muslims yearn to visit Mecca. Pelee is a narrow triangular peninsula of hardwood forest, red cedar savannah, and freshwater marshes, jutting out from the shoreline of Lake Erie, in southern Ontario. Here, in the last half of May, wood warblers (Family Parulidae) flutter from the skies like flowers on the wing. On a good day, you can feast on the colors of twenty-five to thirty different species of warblers—the rich chestnut on the cap and neck of a bay-breasted warbler (*Dendroica castanea*), the flamboyant orange throat of a blackburnian warbler (*D. fusca*), the bold black mask of a common yellowthroat (*Geothlypis trichas*), or the stylish gray-and-lemon patchwork on the wings and tail of a yellow-rumped warbler (*D. coronata*). As beautifully feathered as warblers are, it is their ceaseless energy that I find the most captivating to watch. Pioneer ornithologist Frank Chapman described it best in his 1907 book *The Warblers of North America*. "[They] pirouette or flutter, turning the whole body this way and then that, darting or springing here and

The tiny blackpoll warbler may migrate over seven thousand kilometers to reach its wintering grounds in South America.

there, the embodiment of perpetual motion."

For most of the warblers that alight in Point Pelee, the park is merely a rest stop on their northward migration to the boreal forest. In the taiga, these delicate little flames of life will breed and fatten; they then flee south again, to await another May. By late summer, with an infusion of new recruits, warbler numbers are at their highest. Even so, the autumn warbler migration fails to stir the heart of humans as it does in spring. In autumn, the tiny songsters are silent and most have molted their courtship finery, replacing it with a drab mix of olive, gray, buff, and brown-colored plumage that torments every bird-watcher. But many of these "confusing fall warblers" simply trade their summer finery for fortitude, and in a moment I'll astound you with some of their autumn feats of flight.

The wood warblers come to the taiga in the first place for the seasonal abundance of food—a seemingly endless banquet of leaf-chewing caterpillars, sap-sucking aphids, spiders, ants, beetles, leafhoppers, plant lice, and flies. Since most wood warblers are similar in size (typically 13 centimeters [5 in.] from the tip of their beak, to the end of their tail), and all have the same slender bill, they eat almost exactly the same foods. To lessen the competition between them, the birds frequently forage in different areas. For example, some wood warblers feed mainly in deciduous trees, while others search in conifers, and still others hunt in bushes or scratch in the litter on the forest floor.

Even within a single balsam fir (*Abies balsamea*), different warblers may hunt in different sections of the tree. For instance, the Cape May warbler (*Dendroica tigrina*) forages most often in the crown of the tree and on the outside branches, whereas the black-throated green warbler (*Dendroica virens*) hunts in the middle of the tree, among the dense inside limbs. In the same tree, an American redstart (*Setophaga ruticilla*) might hawk for flying insects from the

Photo: Brian Small

118

The colorful yellow warbler is one of the most wide-spread wood warblers in the North American taiga.

tips of the outside branches, a black-and-white warbler (*Mniotilta varia*) may search the cracks and crevices of the trunk, while a blackburnian warbler carefully scours the length of a large limb, starting near the trunk, flitting, and hopping outwards. Even though all of this sounds nice and clean, Nature is rarely so precise in its partitioning, and the movements of foraging warblers commonly overlap.

Every warbler species has a distinctive song. During the summer breeding season, territorial males sing frequently to advertise their presence, and not surprisingly, where they forage is where they sing. What is really interesting is that the tree height at which male warblers forage not only determines where the bird sings, but also how it sings. Researchers compared the song frequencies of sixteen different wood warblers and discovered that the songs of treetop species averaged 8,100 hertz, medium-level species averaged 6,207 hertz, and low-level species averaged 3,136 hertz. In other words, warblers that forage in the treetops sing high-pitched songs, whereas those that forage in the understory sing low-pitched songs. This makes good sense. High frequency sounds transmit poorly in dense underbrush, so a male with such a song cannot broadcast his message as far as a rival who has a low frequency call that penetrates vegetation better. In contrast, a high frequency song is the best form of advertising in the tree-tops, where there is less interference from surrounding vegetation.

Most wood warblers arrive on their boreal breeding grounds by the end of May. Within four months, the songsters leave again. Most of them abandon the taiga by mid-September and flee to the tropical forests and mangrove swamps of the Caribbean, Mexico, and Central America. In the tropics, the warblers often travel together in mixed-species flocks, and many include nectar and fruit in their diets, as well as insects. The wood warblers are, in fact, tropical birds that spend less than a third of their year in the North. This tropical ancestry is well illustrated by the Cape May warbler's specially adapted tongue. The bird's tongue is fringed to help it lap up sweet nectar in the tropics, where it must compete with local hummingbirds. It's doubtful that the bird's nectar-feeding tongue is of any benefit for the few months that it feeds on insects in the taiga.

To reach their southern wintering grounds,

The chestnut-sided warbler of eastern North America migrates from the taiga in late August and early September.

many wood warblers must fly non-stop for 1,000 kilometers (620 mi.) or more, navigating in the darkness across the featureless waters of the Gulf of Mexico. Others that migrate to distant Caribbean Islands must face the autumn procession of hurricanes, which annually ravages those shores. These are prodigious feats for birds weighing little more than a ballpoint pen. One wood warbler, the blackpoll (*Dendroica straita*), towers above all the rest in its physical achievements, for it travels farther, longer, and higher than all others.

The blackpoll warbler breeds throughout the taiga of North America, right up to its northern limit at the tree line. Let's follow the autumn migration route of one of these warblers and you will marvel, as I did, at the endurance of these remarkable tiny songbirds. In late August, our miniature migrant leaves the familiar conifers along the Mackenzie River and heads southeast, across the breadth of Canada. For the next month or so, it flies at night, resting for days at a time, and feeding and building the vital fat reserves it will need for the longest leg of its

blackpolls rise into the darkness and head southeast over the immensity of the Atlantic. The birds twitter and chirp to each other as they flap into the unknown. By sunset the following day, the flocks are abreast of Bermuda, flying at 40 kilometers per hour (25 MPH), at an altitude of 1,000 to 2,000 meters (3,280 to 6,560 ft.). The continuous exertion has raised their body temperature to 45° or 46°C (113° or 115°F), within a degree or two of their lethal limit. But there is nowhere for them to land and cool off during their transoceanic trip, so they must fly on.

Sometime after sunrise on the second day, the tiny birds cross the Tropic of Cancer, where they encounter the northeast trade winds. By now, they have traveled 1,500 kilometers (930 mi.), and some have climbed to an altitude of 6,000 meters (19,700 ft.), where the oxygen level is half that at sea level and the temperature, below freezing. The heart, lungs, and blood of migrant birds are well adapted to such rarefied altitudes, and the frigid temperatures they encounter at these heights keep the birds cool and lessen the moisture that evaporates from their lungs. Dehyration is one of the greatest threats they face on extended flights such as these. The birds cannot drink saltwater, and if they land on the ocean to rest, they will surely drown.

The northeast trade winds alter the birds' course and gradually drive them towards the northern shoreline of South America. Finally, 2,500 to 3,500 kilometers (1,550 to 2,175 mi.) later, the exhausted survivors flutter from the skies to rest for the first time in as much as eighty-eight hours. In *Living on the Wind*, author Scott Weidensaul estimates that a black-poll warbler may flap its wings three million times to reach the safety of South America. When it lands, it weighs half of what it did when it left the shores of Canada. Weidensaul jokes, "If a blackpoll warbler were burning gasoline instead of its fat reserves, it could boast getting 720,000 miles to the gallon." Somehow, now that I know these facts, the achievements of human Olympians seem embarrassingly trivial.

journey. By early October, the warbler, having flown 4,200 kilometers (2,610 mi.), reaches Nova Scotia, where others of its kind have gathered to wait. The blackpolls are padded with fat to fuel the flight ahead, many having doubled their body weight. The restless warblers scour the forests, biding their time. They need a storm and the northwest winds that normally follow it to proceed with their journey. One night, in the middle of October, the conditions are just right. Between 8:00 PM and midnight, thousands of

Boreal Bruins

I scribbled these bear observations into my journal one September day in northern Ontario.

The bear's nose was never still, twisting from side to side, distorting its massive face with comic effect. Its snout stretched upwards into the moist autumn air, teasing out my scent from the background odors of spruce, damp earth, and decaying leaves. The animal stripped the frozen berries from one more branch, then turned and disappeared into the tangle.

In North America, the most powerful carnivore in the taiga is the bear. In the central and eastern third of the boreal forest, the local bear is the American black bear (*Ursus americanus*). Adult males weigh from 60 to 140 kilograms (132 to 309 lb.), and adult females weigh from 40 to 70 kilograms (88 to 154 lb.). However, when food is plentiful, these ebony bruins can really bulk up. There are records of black bears from the taiga of Quebec and Newfoundland that weighed over 275 kilograms (606 lb.), and in Dryden, Ontario, one adult male tipped the scales at an incredible 329 kilograms (725 lb.)!

In the boreal forests of western North America, the black bear shares the taiga with the brown bear (*Ursus arctos*), although it's not an even partnership. Brown bears, also called grizzlies, are usually larger than black bears and dominant wherever the two species overlap. In fact, brown bears will kill black bears, given half a chance. One October in west-central Alberta, a large female grizzly with two yearlings followed the tracks of a mother black bear and her two cubs through the snow to the family's winter den. The grizzlies killed and ate the two black bear cubs, but the mother bear managed to escape.

Brown bears range throughout the taiga of Eurasia, as well as in the forests of western

The brown bear, or grizzly, is found throughout the taiga of Russia as well as in western North America.

North America. Within this expansive area, they occupy many different habitats, from lush coastal areas, to sparsely vegetated tundra along the tree line. The average male brown bear—if such a creature exists—weighs between 150 and 380 kilograms (330 and 838 lb.), and females weigh half as much.

A bear's size depends mainly on what it eats. Brown bears that feed on salmon in coastal areas are the largest of their kind. Males commonly weigh over 500 kilograms (1,100 lb.), with exceptional individuals exceeding 680 kilograms (1,500 lb.). If one of these huge bears were to rear up on its hind legs immediately in front of you, the animal's massive size would instantly focus your thoughts on the afterlife. Throughout the circumpolar taiga, brown bears and salmon come together only in the Kamchatka Peninsula of southeastern Siberia.

For black bears and browns alike, autumn is the critical season to pack on the pounds and build up the fat reserves they will need to survive their winter denning period. All bears in the taiga den for at least six months every year. The record is held by a female black bear from Alaska who was down and out for 247 days, or just over eight months. While denning, a bear loses from 15 to 25 percent of its body weight. For pregnant females, the energy drain is even greater; mothers with newborn cubs lose as much as 40 percent of their body weight.

Bears are experts at finding the most nutritious foods available at any given time. Their survival depends upon it. In September, they gulp down insects, especially ants (Family Formicidae) and wasps (Family Vespidae), and any carrion they sniff on the wind. They also prey on injured moose (*Alces alces*) calves and even tackle antlered adult bulls that have been wounded or weakened by the rigors of the autumn rut. Bears are stealthy, fast, powerful predators, and they stalk and kill healthy prey as well. In Newfoundland, researcher Shane Mahoney reported that black bears sometimes kill 15 percent of caribou (*Rangifer tarandus*) calves and as many as 45 percent of moose calves every summer. More surprising is the discovery that Newfoundland black bears are the culprits in 46 percent of the annual adult caribou deaths. Mahoney eloquently describes a bear attack on an adult female caribou.

The charge was awe inspiring. The cumbrous clown became an ebony laser, and the 30-meter distance was covered in a few brief seconds. ... [The caribou] made three strides and the bear was on her, his massive forearms bearing the leverage he needed. His claws held her haunches and she was flipped on her side like a doll. His jaws clamped shut her windpipe and she suffocated quickly; her calf ran in erratic circles [around them] and was clubbed lifeless. In 48 hours only the hides and hooves remained.

Although boreal bears are capable predators, 80 to 90 percent of their diet consists of green vegetation, flowers, roots, and berries. In the taiga, bears gobble down dozens of kinds of berries to satiate their autumn appetites. They commonly eat crowberries (*Empetrum nigrum*), bearberries (*Arctostaphylos* spp.), blueberries (*Vaccinium* spp.), and soapberries (*Shepherdia canadensis*). An inquisitive biologist determined that a hungry bruin may eat two hundred thousand soapberries in a single day. Imagine for a moment the dedicated researcher, hunched over a lab table, teasing apart a collection of bear turds and carefully counting the seeds. Now that's dedication to science.

Black bears communicate using vocalizations, head position, body posture, and ear position.

123

Bears such as this denning black bear are the only mammals that give birth while they are hibernating.

Anyone who examines a bear dropping in autumn (which I know most of you do) knows that bears are clean feeders. They ingest very few leaves or twigs when they pluck berries, and the fruit is devoured with a minimum of chewing. There may be a good reason for this. Plants produce berries to entice wildlife to eat them. Later, when the seeds are passed, the plants benefit by having their seeds dispersed. Obviously the plants would not benefit if their seeds were chewed and destroyed during ingestion. To prevent this, many seeds are impregnated with distasteful toxins, some of which can even poison an animal.

Not only do plants benefit from having their seeds scattered by bears, but the seeds also do better at germinating after they've passed through the animal's digestive tract. For wild raspberry seeds (*Rubus idaeus*) collected from black bear droppings, the germination rate is double that for seeds taken directly from the plant. For chokecherry seeds (*Prunus virginiana*), the germination rate is tripled, and for red-osier dogwood seeds (*Cornus stolonifera*), the rate is seven times higher. Researchers believe that the corrosive action of stomach acids and the mechanical abrasion of the seeds during their passage through a bear's digestive tract make the seed coats more permeable to water and gases, thus dramatically improving their germination rate.

Bears maintain their weight throughout the summer, eating between five and eight thousand calories a day (most humans eat between two and three thousand calories daily). Then, in early autumn, a bear's appetite suddenly switches into overdrive. The hungry bruins embark on a feeding frenzy, spending as many as twenty hours a day gulping down from fifteen to twenty thousand calories, the equivalent of chowing down forty-three hamburgers and twelve orders of French fries in one day. Whoever coined the expression "hungry as a bear" must have observed these animals in autumn. When food is plentiful, a bear may gain 0.7 to 1 kilogram (1.5 to 2.2 lb.) a day, eventually increasing its body weight by a quarter or more.

The autumn appetite of a bear appears to be under hormonal control. If food is plentiful, the animal stops feeding once its fat reserves reach the right level. In female bears, added factors, such as pregnancy and lactation, also influence their final weight gain. A female bear needs many calories to sustain a pregnancy and to nurse her newborn cubs while the family is sequestered inside their winter den. In one autumn black bear study, a group of pregnant females gained an average weight of 40 kilograms (88 lb.) each. The following autumn, this same group of females, now nursing eight-month-old cubs, gained just 11 kilograms (24 lb.) each. Keep in mind that black bear cubs stop nursing in their second winter and hibernate along with their mother, so the energy drain on her is much less than when she is pregnant. The following year, after the family splits up and the cubs have left, the mother bear, who is usually pregnant once more, gains an average of 40 kilograms (88 lb.).

The fine tuning of this system of weight gain for female bears continues in a remarkable way, with the size of her fat reserves ultimately controlling the fate of her pregnancy. After bears mate in June, the fertilized egg immediately begins to divide and grows into a hollow ball of cells, the size of a pinhead, called a *blastocyst*. Then, the process suddenly stops. For the next five months or so, the pregnancy is put on hold,

and the blastocyst simply floats in the cavity of the female's uterus. Then, in late November or early December, the pregnancy suddenly starts up again, the blastocyst implants in the wall of the uterus, and the pregnancy proceeds to completion. This pattern of reproduction is appropriately called "delayed implantation." In bears, implantation of the blastocyst occurs *only* if the female has stored enough fat during her fall feeding frenzy to sustain the energy demands of a pregnancy. If the female is too thin, she simply aborts the blastocyst, terminates the pregnancy, and suffers no ill effects. In this way, pregnancy only proceeds when the nutritional condition of the female bear is ideal.

Biologists speculate that female bears may also benefit from delayed implantation because it is a practical strategy that allows mating to occur early in the summer, when feeding is not so critical. If bears were to mate in the fall, breeding activities would seriously disrupt this important feeding period, making it harder for the animals to accumulate the fat reserves they need to sustain them through winter hibernation.

The suspected signal for blastocyst implantation is a simple and reliable one—the number of hours of daylight, or photoperiod. In late autumn, the photoperiod gradually decreases as the days grow shorter. When it reaches a critical threshold, it signals the bear's brain to release the hormones that restart pregnancy and initiate implantation.

In summary, the size of a mother bear's fat reserves determines *whether* she implants or not, whereas the number of hours of daylight determine *when* implantation occurs. These are remarkable refinements in the reproductive biology of bears that illustrate how complex and interesting their lives are, and how sensitively attuned these animals are to their environment.

A Prickly Affair

Everyone knows the punch line to the old joke, "How does a porcupine make love?" Carefully! But courtship and mating between prickly porcupines (*Erethizon dorsatum*) is more than a careful affair. It is a fascinating series of events in which the truth is more extraordinary than anything you could imagine.

September to November is the season of sex for the porcupine. During this time, females are in heat for a number of weeks, but they are only receptive to mating for eight to ten hours in that time. In fact, a thick membrane completely seals the female's vagina for most of the year, except at the peak of the breeding season and while she gives birth. During these two brief periods, the membrane thins and disappears. This protective vaginal membrane likely evolved in the female to prevent stray chips of bark from lodging in her genital tract as she scales and descends tree trunks.

Porcupines are solitary animals for much of the year, so males must make an effort to track down receptive females. They locate them, in part, by the characteristic odor of the female's urine, which she dribbles on the ground, and on limbs and tree trunks throughout her home range. Most sexually receptive females grunt and whine a lot, which also helps the males to find them. Once a male locates a lady-on-a-limb, he climbs the tree with her and waits patiently on a branch nearby. Often he stations himself on a lower limb, where he can guard her from other suitors. During his vigil, he periodically hauls himself up to the female and sniffs her body to test her receptiveness. If his timing is not quite right, the female may bite him, or worse, slap him in the snout with her tail. Her suitor then retreats to another limb to pull the quills out of his face and bide his time.

Since the female is in heat for several weeks, all the adult males in the neighborhood soon learn of her sexual condition and come shuffling. In this way, she insures that the largest and most vigorous males vie for her attention. Several males may gather in the same tree as a female, and males of equal size may actually battle each other.

Twenty-five years ago, I observed a fascinating porcupine *ménage à trois* in Ontario. Our encounter started when I heard a loud whining call from a cluster of trees, 50 meters (55 yd.) away from where I was hiking. I had never heard such a strange call before and had no idea what it was. I certainly never expected the caller to be a porcupine. As I approached, I discovered there were three porcupines in the same tree. One animal was huddled on an upper limb; the two others were on a lower branch, facing each other in a standoff, with one doing all the whining. On the ground around the base of the tree, I noticed many discarded quills and some small tufts of black hair. Later I learned that the two combatants had probably been males, fighting over a female in heat.

During mating battles, male porcupines are sometimes knocked out of their tree. It seems unusual that such accidents should happen to an animal that spends so much of its life climbing aloft in trees. However, a study in New York state revealed that a third of the porcupines there had healed fractures, presumably sustained in such falls. When a porcupine falls out of a tree, it may not only fracture a rib, leg, or hip, but it may also impale itself on its own quills. To protect itself from the consequences of such punctures, a porcupine's quills have evolved a safety feature—they are coated with grease that has antibacterial qualities. The greasy coating prevents infection from developing when a quill becomes imbedded in flesh and cannot be extracted. This antimicrobial grease also minimizes inflammatory reactions in other animals that might be stabbed with a quill, including inquisitive biologists. When veteran researcher Dr. Uldis Roze tried to capture a 8-kilogram (17.6-lb.) adult male, the angry animal drove a quill into the

*A porcupine feeds
on rose hips stick-
ing above the snow
in -20°C weather.*

inside of his upper arm. The quill emerged two days later from Roze's forearm, 25 centimeters (10 in.) below the point of entry. Surprisingly, he admitted that the embedded quill had produced a minimum of discomfort.

When a female porcupine finally wants to mate, the deed usually occurs on the ground. The last stages of courtship are quite remarkable. Typically, the successful male suitor waddles towards the female on his hind legs, bracing himself with his tail. He grunts repeatedly, as he shuffles closer and closer. When he is about half

a meter (1.5 ft.) away from his love of the moment, his penis suddenly springs to life, and he completely soaks her with blasts of urine. I presume his intentions are to whet her appetite and not to dampen her spirits. After the dousing, a chase or two, some wrestling, and more showers of urine, the wooing male carefully mounts his intended. His female partner lessens the risks of such a coupling by arching her tail over her back. The underside of a porcupine's tail is covered with firmly anchored bristles, not detachable quills, so the male can safely lean his

Young porcupines finally separate from their mother in late summer.

chest on her. Sexual contact lasts only two to four minutes, but repeated copulations may prolong the affair for several hours. One vigorous male copulated eight times in just twenty minutes. The observer commented that the male "seemed completely exhausted afterwards." That's a surprise.

Eventually, when the coupling is over, one of the pair usually waddles off and climbs a tree. If its partner follows with further amorous intent, the pursued one screams its displeasure. The rejected lover has no recourse but to trot off and resign itself to being loved and left. Within an hour of mating, a visible mucous plug forms in the female's vagina—a kind of chastity belt—which prevents any further mating attempts.

Porcupines are found everywhere in the taiga of North America, except Newfoundland. These well-armed rodents are active throughout the winter. They spend the daylight hours hidden in a rocky crevice or tree cavity and emerge at night to gnaw on the sapwood of trees. They feed on the inner bark of every tree species in the taiga, as well as on the bark of willows (*Salix* spp.) and alders (*Alnus* spp.). Porcupines are not well designed to plow through deep snow, so they usually den as close as possible to the trees they like to eat. In fact, during most winters

they travel less than 100 meters (328 ft.) from their daytime den to their nighttime feeding trees, in an attempt to minimize the energy they burn. Nonetheless, porcupines in the taiga may lose a quarter of their body weight during the demanding winter season. Dr. Roze summarized the animal's plight in his engaging book *The North American Porcupine*. "During winter, porcupines slowly starve to death—their weight drops steadily, their fat reserves are depleted, and they reach a low point of their existence."

Most porcupines are born sometime in May. The animals' lengthy gestation period of 210 days is double that of the brawny beaver, which is three times larger in size. Also, porcupines give birth to a single, large offspring, whereas beavers normally have three or four small kits, and as many as nine. The young arboreal porcupine, called a *porcupette*, is much more precocious than its aquatic relatives. Porcupettes are born with their eyes open, their teeth erupted, and their bodies protected by splinter-sized quills that harden within an hour of birth. These newborns can flail their tails in self-defense and climb before they are two days old. If you are a plodding, rotund, shortsighted rodent that feeds in trees, your youngster needs to be up and running from the moment it's born.

Wolf Farts and Witch's Butter

Fungi are unusual organisms, indeed. Although they look like plants, they have no stems, leaves, or roots. Fungi also lack chlorophyll, so they are unable to photosynthesize food as plants do and must scrounge all of their nutrients from either plants or animals, alive or dead. Fungi, in fact, are neither plant nor animal and are so different from both of these groups of organisms that they comprise their own unique kingdom, the Fungal Kingdom (Kingdom Mycota).

Fungi can be found throughout the year, but the greatest diversity of species appears in early spring and early autumn. The warmth of spring, together with the moisture from melted snow, stimulates early growth, but it is short-lived. By late spring, fungi numbers drop to their lowest point. They slowly increase again throughout the summer, and then, if the chill of early autumn (usually September) is accompanied by rains, they reach their highest point of the year, enlivening the forest floor with their variety and color.

One species of fungus or another eventually permeates every woody fiber of the boreal forest. Hidden from view most of the time, fungi slowly recycle, dissolve, digest, and transfer nutrients from one life form to another. As you stroll through the forest, you may see mushrooms on the ground and on tree trunks; these are the fungal equivalent of flowers. Each of these fruiting bodies releases millions, and sometimes trillions, of microscopic spores, which enable fungi to spread far and wide. Lofted by the wind, the dustlike fungal spores can be transported thousands of kilometers from their source. Each spore is coated with a thick cell wall that resists drying and helps it survive for years after it settles.

A small number of boreal fungi employ more imaginative methods than the wind to disperse their seedlike spores. Earthstars (*Geastrum* spp.) and puffballs (*Lycoperdon* spp.), for exam-

ple, rely on the gentle impact of raindrops to release tiny puffs of spores. The club-shaped, stinkhorn fungus (*Phallogaster saccatus*), which occurs only along the southeastern boundary of the North American taiga, scatters its spores in the most creative way of all. When mature, the stinkhorn smells like rotting flesh. The fetid odor attracts carrion-feeding flies and other insects, which devour the green slime that

Many boreal animals, including red squirrels, northern flying squirrels, red-backed voles, and deer mice, eat fungi.

covers the cap of the mushroom, and in the process, are dusted with spores.

Buried beneath the colorful, sometimes fanciful, mushrooms that erupt in the forest hides the main body of the fungal organism—an interwoven mesh of threadlike filaments called *hyphae*. The hyphae are microscopic in size, and a thousand strands may be no thicker than a human hair. Incalculable kilometers of these

Woody, shelf-like fungi, such as this species, may live for many years, adding a new ring of growth each year.

One species of fungus or another eventually permeates every woody fiber of the boreal forest. Hidden from view most of the time, fungi slowly recycle, dissolve, digest, and transfer nutrients from one life form to another.

fibers. Bark normally prevents invasive fungi from penetrating wood. In fact, healthy bark produces toxins that deter fungal growth, and as long as the bark remains intact, the tree is protected. In the taiga, however, many forces can wound a tree and leave it open to fungal attack and penetration: the scrapes of adult bull moose (*Alces alces*) and caribou (*Rangifer tarandus*), thrashing the velvet from their antlers; the slash of claws when a black bear (*Ursus americanus*) climbs aloft to escape the torment of biting insects; the charred scar from an incandescent bolt of lightning; the gnawings of a hungry beaver (*Castor canadensis*), red squirrel (*Tamiasciurus hudsonicus*), or snowshoe hare (*Lepus americanus*); or the abrasions and broken limbs that result from colliding with a neighboring tree, weakened by senescence and toppled by the wind.

Once tree bark is lacerated, fungi soon invade. The woody decay of living trees is a blessing for wildlife. Chickadees (*Parus* spp., *Poecile* spp.) and many taiga woodpeckers (Family Picidae) build their nests in the softened, friable wood of these afflicted trees. In fact, pileated woodpeckers (*Dryocopus pileatus*) target aspens (*Populus tremuloides*) that have been invaded by heartwood fungus rot (*Phellinus tremulae*). Each bird may excavate three or four cavities, sometimes up to ten, which it uses for nesting and roosting.

I once spent an engaging day with researcher Dr. Rick Bonar in northern Alberta and learned that "pilos" are considered keystone species that play a vital role in the ecology of the forest. Abandoned pileated woodpecker cavities are crucial for many critters in the taiga, including northern flying squirrels (*Glaucomys sabrinus*), red squirrels, common goldeneye ducks (*Bucephala clangula*), American kestrels (*Falco sparverius*), as well as three species of small owls, the northern pygmy owl (*Glaucidium gnoma*), the northern saw-whet owl (*Aegolius acadicus*), and the boreal owl (*A. funereus*). Woodpecker holes also weaken a tree's trunk, so the tree may snap in a windstorm, leaving behind a broken snag. Such snags are favored

fungal threads infiltrate the soil, the roots, and the wood of their hosts. Here, veiled from the sun, most hyphae work in one of three ways: as parasites, saprophytes, or mycorrhiza.

Parasitic fungi infect *living* hosts, causing decay. Some attack and digest lignin, whereas others dissolve cellulose; both lignin and cellulose are essential components of woody tree

Fungi in the boreal forest are often most varied and abundant in the forests of early autumn.

Photo: J. F. Bergeron

nesting sites of the northern hawk owl (*Surnia ulula*), the great gray owl (*Strix nebulosa*), and the barred owl (*Strix varia*).

Saprophytic fungi dissolve and decompose *dead* animals and plants. In most ecosystems worldwide, fungal saprophytes are essential in the breakdown and recycling of nutrients, especially in the moist, sultry tropics. However, in the cold northern forests of the taiga, fungi share the role of recycling with fire, and there, fire is probably the more important of the two processes.

Recently, scientists have learned to use the digestive action of saprophytic fungi to benefit the environment. In the pulp and paper industry, strong poisonous chemicals are normally used to soften wood chips and to separate the fibers so that they can be pressed into a sheet of paper. The noxious chemicals they use are treated and poured into rivers, a dangerous and costly procedure. Some saprophytic fungi, called white rot fungi, are now being tested and used to pretreat wood chips and soften them so that fewer harmful chemicals are used and ultimately dumped into the environment.

Mycorrhizal fungi are the ones that interest me most, and it's only in the past twenty years that scientists have appreciated how important they are to the health of the forest. The hyphae of mycorrhizal fungi penetrate the fine root hairs of at least 90 percent of the tree species that grow in the boreal forest. The relationship between the mycorrhiza and plant roots appears to benefit both partners. The fine branching mesh of the hyphae greatly increases the surface area through which a tree can absorb moisture and minerals. In exchange, the tree supplies the fungus with nutrients, especially sugars that it cannot manufacture itself because it is unable to photosynthesize. As much as a quarter of all the sugars produced by a tree may be diverted to its mycorrhizal partner, buried in the soil below. Trees lacking mycorrhizal fungi grow poorly and are weak competitors. It seems that nutrient siphoning by the fungus is a necessary cost of survival for trees.

Everyone has eaten commercial mushrooms.

I'm told they are grown in darkened buildings, in trays filled with shredded manure. This fact alone may be what nudges some people to harvest their own mushrooms from the wild. However, I've never been able to make that leap. The risk of intractable vomiting or hallucinations after a meal of wild mushrooms has never held much appeal for me. Author Eugene Bossenmaier, in his book *Mushrooms of the Boreal Forest*, lists roughly two hundred species of mushrooms, of which 20 percent are edible. His other categories include: unknown edibility (a quarter of the species described); inedible; edible, but not recommended because of their

repulsive odor, acrid taste, or resemblance to poisonous species; edible with caution, meaning you might experience unpleasant side effects; poisonous; and deadly. Forget it. Manure or no manure, I'm getting all my mushrooms from Safeway.

Poisonous mushrooms contain toxins. In small doses, some of these toxins have hallucinating and intoxicating side effects, and for this reason, these so-called psychedelic mushrooms have been ingested by humans for centuries. In *Alice's Adventures in Wonderland*, a talking caterpillar convinces young Alice to nibble on a mushroom, and the heroine promptly shrinks to

The edible fungus called witch's butter grows on the trunks of dead aspens and spruce trees.

Mushrooms such as this red brittle cap grow in great numbers in some boreal habitats.

The urine seems to be more powerful than the mushroom, and its effects may last through the fourth or fifth man.

Those Russians sure knew how to party!

Wolf farts and witch's butter, mentioned in the title of this section, refer to two other taiga mushrooms, but neither will give you the psychedelic trip of a lifetime. Wolf fart is the literal translation of the scientific name of puffball mushrooms (*Lycoperdon* spp.). Perhaps the name refers to the musty brown puffs of spores that spout from the top of a ripe specimen. Or perhaps it has some other, more Freudian meaning, but I've never felt it was safe to delve too deeply into the mind of a mycologist (someone who studies fungi). Witch's butter is a purely descriptive common name for *Tremella mesenterica,* a bright yellow, jelly-like fungus that grows on the trunks of spruce and poplars. Apparently the fungus is edible, but you won't catch me spreading it on my toast.

Any discussion of fungi would not be complete without a brief word about lichens. Worldwide, there are about twenty thousand different lichens, of which several thousand grow in the boreal forest, making them one of the most diverse forms of life in the taiga. Earlier in the book I talked about reindeer lichens (*Cladonia* spp.), most of which grow as miniature clubs or multi-branched clusters. Many other boreal lichens simply grow as colorful crusts on bark, rocks, weathered metal, discarded antlers, and soil.

When I was in high school, I learned that a lichen was a partnership between a fungus and an alga. The fungus provides a moist refuge where chlorophyll-rich alga can photosynthesize and produce sugars for both organisms to share. Together, they live happily ever after. Modern research indicates that algal cells, which can grow perfectly well on their own, are actually prisoners of the fungus. One imaginative lichenologist described the organism as the "union between a captive algal damsel and a tyrant fungal master."

less than 30 centimeters (1 ft.) tall. The moral of the story is clear. Never listen to a caterpillar, especially one that smiles at you.

One toxic mushroom of the taiga, fly agaric (*Amanita muscaria*), has received more publicity than any other. In 1774, the famous naturalist Georg Wilhelm Steller reported that the Natives of the Kamchatka Peninsula, in Siberia, routinely ate these hallucinogenic mushrooms in small amounts, especially during feasts and celebrations. He wrote in his journal:

> After about half an hour the person becomes completely intoxicated and experiences extraordinary visions. ... Those who cannot afford the fairly high price, drink the urine of those who have eaten it, whereupon they become as intoxicated, if not more so.

Hoarding for Winter

The North American red squirrel (*Tamia-sciurus hudsonicus*) is one of the most conspicuous animals in the boreal forest, not because it is particularly visible, but because it is such a loudmouth. Walk through almost any grove of white spruce (*Picea glauca*) in the taiga, and a hot-tempered squirrel will boldly chatter at you for trespassing. The red squirrel is the most vociferous of the tree squirrels, which explains some of its nicknames, such as barking squirrel, boomer, and chatterbox. A similar species of red squirrel (*Sciurus vulgaris*) also ranges across the taiga of Eurasia.

When the squirrel is not scolding some intruder, it's racing through the treetops, busy with life. The legendary naturalist Ernest Thompson Seton thought the squirrel "bounces about among the branches like an animated rubber ball, moving at such a rate that it becomes a blur." Searching for food, eating food,

In Alaska, an industrious red squirrel may cache up to 16,000 white spruce cones in a two-month period.

This mushroom was wedged in a black spruce by a red squirrel that later forgot to retrieve it.

One day, whilst visiting the house, I was called outside and here was the squirrel laboriously dragging a chicken as big as itself by the neck up a tree. On looking more closely, two other chickens were discovered, hung by their heads in forked branches. The three dead chickens had all been killed by bites at the back of the head.

Needless to say, the gun-wielding farmer was not amused. Moments afterwards, the squirrel was on his way to that big spruce forest in the sky.

In late summer, the red squirrel begins to hoard for winter. The cached food helps it survive the months of winter cold and privation. Each individual squirrel defends a small home territory. Since trespassers might steal food from a cache, the red squirrel aggressively chases all intruders, including its own offspring, and drives them out of its territory.

Animals hoard food in two different ways—they scatter it or clump it. Gray jays (*Perisoreus canadensis*), ravens (*Corvus corax*), crows (*Corvus brachyrhyncos*), and chickadees (*Parus* spp., *Poecile* spp.), for example, are scatter hoarders, hiding one or two items of food in numerous locations. In contrast, the beaver (*Castor canadensis*) stashes its winter store of branches in a single, large underwater pile, next to its lodge. The red squirrel hoards in both ways.

Every squirrel maintains at least one primary midden, a surface pile of discarded cone scales and debris, mixed with old and fresh conifer cones. It may also have several smaller, satellite middens in the core of its home territory. Middens are easy to spot because of their large size; some can be 10 meters (33 ft.) in diameter and 40 centimeters (16 in.) deep. Sometimes red squirrels cache food in numerous scattered locations, in addition to stockpiling it in a central midden. In a lodgepole pine (*Pinus contorta*) forest, in the Cypress Hills of southeastern Alberta, red squirrels cached roughly the same number of cones in their primary midden as they scattered throughout the forest nearby. In scattering their cones, the average squirrel traveled an estimated

transporting food, and caching food keeps the red squirrel in seemingly perpetual motion. In summer, the animal gnaws on mushrooms, berries, seeds, and tree buds, but it also has a taste for meat. Red squirrels eat the eggs and nestlings of birds, and in Manitoba, one was seen hauling a dead tiger salamander (*Ambystoma tigrinum*) up a tree. It's one thing to prey on helpless nestlings and sluggish amphibians, and quite another to kill hot-blooded mammals, but squirrels are surprisingly capable of doing this as well. A 1993 report from Yukon revealed that two-thirds of newborn snowshoe hares (*Lepus americanus*), called *leverets*, died in their first two weeks of life. Half of these mortalities were caused by killer red squirrels, which often carried the dead leverets into treetops, to stash them for a later meal.

Sometimes a squirrel's carnivory can cause it trouble. Writer Reginald Buller described what once happened at a farmhouse in southern Manitoba.

32 kilometers (20 mi.). This is a big investment of energy, so there must be a definite survival advantage. It's unknown whether red squirrels in the boreal forest adopt a similar mixed-caching strategy.

Throughout the taiga, red squirrels thrive best in dense stands of spruce that shade the understory and keep their middens cool and damp. Cones that are stored under these conditions do not open, and their seeds remain viable and edible for years afterwards.

When red squirrels begin to stash surplus food for the winter in late summer, they race through the treetops, nipping off cones and letting them fall to the ground. Later, the squirrels gather up the fallen cones and bury them in middens. Their favorite cones are those of the white spruce, and an industrious squirrel may gather a thousand of these cones in a single day.

In autumn, a red squirrel typically spends 80 percent of its day cutting and hoarding conifer cones, and it may cache up to fourteen thousand spruce cones in a six to eight week period. However, conifer cones are not the only food stored by squirrels. They also cache cranberries (*Vaccinium* spp.), bearberries (*Arctostaphylos* spp.), and crowberries (*Empetrum nigrum*), as well as eighty-nine different kinds of mushrooms. In fact, the red squirrel eats a greater variety of mushrooms than any other boreal mammal, including voles (*Microtus* spp.), mice (Order Rodentia), lemmings (Family Muridae), shrews (*Sorex* spp., *Blarina* spp.), and flying squirrels (*Glaucomys sabrinus*).

Mushrooms are not simply plucked and stored by squirrels. First, they are dried to prevent decay, and squirrels do this quite methodically. A squirrel may try three or four different locations before it finds just the right spot to dry a mushroom. It may even change trees in the process. The ideal drying site seems to be the outer half of a bushy spruce limb, from 1 to 10 meters (3 to 33 ft.) off the ground. Squirrels seem to select the sunny side of a tree more often than the shaded side, but I must admit I haven't done the science to prove this. I love to wander through a squirrel's territory in September and see how many mushrooms I can find wedged along the branches of the spruce trees. My record is thirty-six.

Mushrooms are 70 to 90 percent water and once they shrivel up, they can be stored in a relatively small space. As far as I know, red squirrels do not store mushrooms in moist middens, where they would surely rot. More often, they store mushrooms in tree cavities or in abandoned woodpecker nests, and researchers have found as many as three hundred mushrooms in a single stash.

Some of the mushrooms that red squirrels cache and eat are poisonous to humans, including fly agaric (*Amanita muscaria*). Maybe it's the ingestion of these hallucinogenic mushrooms that makes the squirrels bounce off the branches? These animals eat poisonous mushrooms with apparent impunity, but some authors have suggested that the toxins might accumulate in squirrels' tissues. In their celebrated book *Up North,* trivia buffs Doug Bennett and Tim Tiner claim, "Woodlore traces the expression 'squirrelly' to trappers and others who went a little funny eating too many squirrels, suggesting active ingredients remain potent in their flesh." I don't know whether this is true or not, but I've met my share of trappers, and more than a few of them were "a little funny" for reasons other than their appetite for roasted red squirrels.

Bulls in Rut

For me, the bull moose (*Alces alces*) is the most impressive animal in the boreal forest. In northern Ontario, I heard many stories about hormone-charged bulls challenging locomotives, and losing. One of my favorite stories involves three urban hunters who came north to "bag themselves a trophy bull." As the trio drove down a remote logging road in their shiny new truck, a foul-tempered bull stomped onto the road and rammed the front of their vehicle. The three terrified, rifle-packing tough guys cowered in their custom leather seats. The bull eventually lost interest in the ramming match and ran into the bush to look for a more responsive rival. That was the good news. The bad news was that the animal had dented the fenders so badly that the vehicle could no longer be driven. After an hour or so, the petrified hunters finally plucked up the courage to abandon their disabled truck and run home to the safety of civilization. They learned a valuable lesson that day: some wildlife really is wild.

In September 1999, I spent a memorable five hours with a bull moose and a cow he was shadowing. I wrote these words in my notepad.

It's 10:17 AM. In the gray cast of early morning, the forest is indescribably beautiful. A light snowfall has edged the black branches of the spruce trees with a rim of white. I have been sitting alone for the past two hours, only 15 meters (50 ft.) from a pair of moose that are quietly lying in the still of the trees. A strong musky odor hangs in the moist air; perhaps it's the scent of masculine urine, advertising the bull's health and vigor, or the bouquet of a cow in heat. The female moose is fast asleep, her head stretched out fully on the soft snowy carpet of the forest floor. The only sound is the gurgle of the cow's belly as she ferments her last meal in the watery vat of her rumen. The bull is lying about 10 meters (33 ft.) from his con-

sort. He is expressionless, and his eyelids are heavy. Occasionally, fluffy lumps of soft snow fall from the branches overhead and splatter silently in the bowl of his antlers.

10:48 AM. The bull is losing his fight with fatigue. His head slowly begins to droop. Just as his antlers are about to strike the ground, he jerks awake; but the sequence repeats itself half a dozen times. Finally he snoozes, the tip of his left antler resting on the sphagnum.

12:30 PM. The cow suddenly rises, blinks at me, and wanders deeper into the forest. The weary bull follows, but never closer than 10 to 15 meters (33 to 49 ft.). As he plods along, he grunts, slobbers, and laps his tongue noisily. The pendulous bell, dangling beneath his throat, is soaked with drool. The cow stops to strip some yellowing leaves from a cluster of willows and graze on some dry grasses. The bull also eats a mouthful or two, but mostly he just stands and waits patiently. Twice, the big male vigorously thrashes some bushes with his polished rack. Perhaps he is talking to me, to some unseen antlered-rival, or to his prospective mate? After the second session of bush-battering, the female urinates. The bull grunts, then buries his nose in the steaming moisture in the grass. As he slowly lifts his head, he curls his upper lip and grimaces with the silly face biologists call "flehmen." For him, this is no silly game. It may be his first time to mate, his last, or his only time. The taiga ensures that only the strong and the persevering live long enough to breed.

The mating season for moose is surprisingly synchronous across the breadth of the taiga; it happens in autumn, from the end of September into the first third of October. In fact, 89 percent of cow moose conceive within a ten-day period. Breeding synchronization also occurs in caribou

This bull moose recently rubbed the velvet from its antlers, which are still stained with dried blood.

(*Rangifer tarandus*). In both ungulates, the simultaneous birth of calves the following spring inundates predators and may lessen their over-all mortality; however, moose calf mortalities may still exceed 80 percent.

Cow moose determine *when* a bull can mate, and in much of the taiga, *where* he can mate. Most cows select a small mating area, some-times no larger than the parking lot of a large shopping mall. The area may include the shore-line of a lake or a few small ponds where the adjacent forest is open for good visibility, allow-ing the cow to monitor the approach of prospec-tive suitors. Once the lady has acquired some real estate, she moans about it. Female moose in heat produce long quavering moans that can be heard 3 kilometers (2 mi.) away. As they move around their mating area, they urinate fre-quently, often in shallow water. Bull moose find the sound of trickling urine irresistible, and human hunters often use this sound to lure bulls closer.

Moose have a gland that empties into the cleft in their hooves. When cows in heat scrape the ground with their front hooves, the secretions from these glands may advertise the animal's sex, age, and possibly its reproductive condition.

The moaning, the urination, and the pawing all serve one end—to attract the biggest, tough-est, most seasoned bulls in the neighborhood. A bull moose reaches its maximum body and antler size between eight and twelve years of age. A prime bull may weigh over 450 kilograms (992 lb.) and carry a massive 27-kilogram (60-lb.) palmate rack that spans nearly 2 meters (6 ft.) from tip to tip. With this kind of might and weaponry, bulls are built for battle. To settle a challenge, rivals first try to intimidate each other, but when that fails, they lock antlers and test each other's strength and endurance. Usually serious fights end after several minutes, but some protracted battles may continue for hours, with the pugilists occasionally resting to regain their strength and resolve. Injuries are

The mating areas of many cow moose commonly include a lakeshore or several small ponds that offer good visibility.

Moose avoid continuous coniferous forest, preferring instead the aspens, willows, and shrubbery that thrive after a forest fire.

common. In an Alaskan study, 10 percent of bulls sustained severe head and facial lacerations. Some were blinded, others fractured a rib or a shoulder blade, and a few had lethal puncture wounds to their chest or abdomen.

The bones of a bull moose may be more vulnerable to fractures than expected. Antlers in their final stage of growth are mineralized with calcium. Bulls demineralize their own ribs and shoulder blades by up to 40 percent to meet the heavy calcium demands of their growing antlers. This may explain why these are the bones bull moose commonly fracture when they fight.

A bull on the trail of a cow in heat is dogged in his persistence. Grunting and drooling, he shadows her movements for days. Hormones in the bull's saliva seem to stimulate a cow. Rutting bulls commonly dig pits with their front hooves, urinate and slobber into them, then wallow and splash the muddy slurry onto their antlers, belly, and neck. The scent of this mixture is mesmerizing for a receptive cow, and she lets him approach her. Soon the bull is nuzzling the cow's groin, licking and testing her urine, and resting his chin on her rump. Shortly afterwards, he mounts her, and it's all over within a matter of seconds. Some cows will mate only once, whereas others mate half a dozen times. After a

bull has copulated with a female, he abandons her to search for another. Because the breeding season of taiga moose is so synchronized, most bulls will have time to mate with just one or two cows before the rut is over.

Many of the photographs of mating moose that you see in books were taken in Alaska's Denali National Park; they often show a bull moose surrounded by as many as eight or ten females. Moose are flexible in their behavior, and the breeding pattern of moose throughout most of the taiga is different than it is in the open treeline forests of Alaska and Yukon. Among the so-called tundra moose of Denali, for example, it's the bulls that select the mating area, not the cows, and they wait for the females to come to them to evaluate their status and desirability. In this system, a prime tundra bull can mate with half a dozen cows or more, whereas in the taiga, the average bull moose mates with just one or two partners.

During the rut, a bull moose may not eat for two weeks or more, and when it's all over, he may have lost 90 kilograms (200 lb.). Weak, exhausted, and often wounded, he now must face the wolves of winter. In late November, his antlers drop. Gone is the fire for passion and taste for battle. Survival is all that matters now.

Fangs in the Forest

The moose (*Alces alces*), the beaver (*Castor canadensis*), and the loon (*Gavia* spp.) may embody the spirit of the taiga, but the true lifeblood of this vast conifer forest is the wolf. One summer, a pack of five white wolves visited my campsite in the Arctic. As they left, one loitered behind, howled, and howled again. The sound was energizing, and I wrote in my field journal: "The haunting howl of a wolf penetrates deep into the soul of the listener, stirring the embers of life and fanning them into glorious flame. The wolves were gone in a matter of minutes, but the sight and sound of those legendary beasts flushed me with excitement for hours afterwards."

In North America, three wild canids hunt in the sunlight and shadows of the autumn taiga: the red fox (*Vulpes vulpes*), the coyote (*Canis latrans*), and the gray or timber wolf (*C. lupus*). The red fox is the most widely distributed carnivore on Earth, and it occurs all across the boreal forests of Eurasia, as does the gray wolf.

The three canids have remarkably similar breeding biology. All of them mate in the cold of January or February and give birth to litters of four to six pups in April or early May. Most of the young, born in earthen dens, are sheltered there

from the inclement weather of early spring and protected from hungry predators. However, black bears (*Ursus americanus*) and grizzlies (*Ursus arctos*) sometimes prey on the young pups, as do wolverines (*Gulo gulo*) and golden eagles (*Aquila chrysaetos*). Large canids also willingly prey on the smaller ones.

Young foxes, coyotes, and wolves first appear at the mouth of the den when they are around four weeks old. By then, their eyes are open, and energy fills their every fiber. Wrestling, chewing, and chasing matches among littermates help them develop coordination and build strong bones. Play serves an important social function as well. A serious pecking order develops between young canids at this age, and skirmishes are unexpectedly vicious. In *Red Fox: The Catlike Canine*, wildlife biologist Dr. J. David Henry gives a vivid description of what happens between three and four weeks of age. "Fox pups do not act like cute cuddly puppies such as those of the domestic dog; rather they have always seemed to me to have a slightly demonic character. They are tough month-old thugs, little street fighters, who initiate fights and establish a strict dominance hierarchy during the following ten days." In coyotes and wolves, a similar struggle for rank occurs at roughly the same age. In all of them, the dominant pups in a litter eat first. When food is scarce, those that are ranked lowest go hungry and sometimes starve.

Beginning in early summer, the families abandon the natal den and move to a succession of temporary sites within their home territory. This may be to escape their flea-infested homesite or to hunt in different parts of their territory and monitor against intruders. During these moves, the pups gradually gain confidence and hone their hunting skills for the days when they will be on their own.

The three boreal canids all commonly defend a home territory and mark their ownership with

A wolf may howl any time of the day, and in any season.

In the North American taiga, up to a quarter of red foxes have unusual coloring; they are called cross foxes.

droppings and splashes of musky urine. Coyotes and wolves also advertise their tenancy with howls that fill the night skies with their wild music.

Families begin to break up by the middle of autumn, and the pups disperse widely. When food is plentiful, the pups of all three species may stay with their parents that first winter and share the family hunting territory. The following spring, the yearlings become "helpers" at the den and assist in raising the next litter of pups. Yearling helpers benefit by gaining valuable experience in rearing pups, but as long as they remain in their parents' home territory, they

In North America, three wild canids
hunt in the sunlight and shadows of the
autumn taiga: the red fox, the coyote,
and the gray or timber wolf.

have little chance to raise pups of their own. In wolves, for example, the alpha or dominant female usually suppresses reproduction in all the subordinate females in the pack. Even when a low-ranking female manages to bear pups, the dominant female may kill the pups or force the mother to abandon them. Among red foxes, helpers are relatively uncommon. In coyotes, one or more pups may stay with their parents for part of the winter, forming packs of four to six animals, but then at the start of the next breeding season, the family breaks up, and the yearlings disperse.

Wolves commonly form packs. The largest pack I know of was spotted in Wood Buffalo National Park and contained forty-two wolves. Normally, a wolf pack contains a dozen animals or less. Hunting as a pack enables wolves to attack and kill adult moose and wood bison (*Bison bison athabascae*), the largest, most powerful animals in the boreal forest. Despite the potential nutritional benefits of living within a pack, the subordinate members must balance these advantages against the loss of reproduction. For this reason, lower-ranking pack members often split off and start packs of their own, if there is suitable habitat and prey available.

Red foxes, coyotes, and wolves may live in the same tract of taiga. To minimize competition, they partition the resources in several ways. One way they avoid each other is by using different areas within the forest. The red fox commonly hunts in the openings along roadsides, power lines, and along the edges of fields and meadows. The adaptable coyote prefers open forest, where tracts of trees are mixed with clearings such as those created by wildfires and logging. In contrast, the wolf thrives best in undisturbed forest.

It's the differences in diet, as much as habitat, that separates the three boreal canids, although there is considerable overlap. All of them are scavengers and will feed on carrion when an opportunity arises. In northern Yukon, I once saw a red fox steal scraps from a dead caribou that a grizzly was guarding. Coyotes often follow the tracks of wolves to scrounge the leftovers from their kills. And one winter, I watched a solitary wolf tear into the bloated carcass of a moose that had drowned after falling through the ice on a river.

The diet of each canid is dictated mainly by its body size. The 4- to 7-kilogram (9- to 15-lb.) red fox is a mouser, hunting voles (*Microtus* spp.) and mice (Order Rodentia), as well as red squirrels (*Tamiasciurus hudsonicus*) and grouse (*Falcipennis* spp., *Bonasa* spp.). This russet-coated carnivore also eats fruits, berries, beetles, and crickets. The snowshoe hare (*Lepus americanus*) is probably at the upper size limit of what a red fox can manage to kill.

In the taiga, the 15- to 18-kilogram (34- to 40-lb.) coyote is a dedicated hare-chaser, but in the words of author Frank Dobie, "the coyote's favorite food is anything he can chew." Along the southern fringe of the taiga, coyote packs may kill healthy adult white-tailed deer (*Odocoileus virginianus*), a prey much larger than they usually tackle, and one more often associated with wolves. Deer-hunting coyotes are most successful when deep winter snow impairs the deers' ability to escape.

The muscular 30- to 70-kilogram (66- to 154-lb.) wolf hunts the large hoofed mammals of the taiga: caribou, moose, and bison. In summer, some will stalk beaver, but that burly rodent is usually the smallest prey a wolf will hunt.

Even when there is no overlap in diet or

Coyote pups may stay with their parents over winter, forming small family packs.

Opposite page: The large ears of the red fox help it to pinpoint the rustle of mice and voles under the snow.

*Displays of the aurora borealis are
more frequent around the time of the
spring and autumn equinoxes.*

habitat, the three canids never live on friendly terms. Coyotes will kill red foxes when they can, and wolves will kill coyotes and red foxes. As a result, when wolf numbers are high in a tract of taiga, coyotes are usually uncommon, and when coyotes are abundant, foxes are scarce. In the mid-1800s, humans waged war against the wolf because everyone believed it was a "beast of waste and desolation." This mindless campaign of extermination created a void in the wilderness, and the adaptable coyote was quick to fill the vacancy. Originally the coyote was the wily song dog of the central Great Plains, but within a hundred years it had conquered the continent, ranging from Alaska to Costa Rica, and from the Atlantic to the Pacific. In the past thirty years, the fortunes of the wolf have begun to rise again. Wolves now inhabit many areas from which they had been previously eliminated, and coyotes have been forced to retreat from center stage in many areas.

I recently traveled to the Northwest Territories to capture a few final images of the northern lights. One night, while I waited alone in the darkness beside a small lake, I heard the distant howl of a solitary wolf. Almost on cue, the sky was filled with swirling curtains of dancing green light. The light faded and brightened, sped up and slowed down, as if the sky was breathing in accompaniment with the wolf. That night, the brilliance of the aurora and the howl of the wolf made me reflect on the impact that humans have on these two natural wonders of the taiga. Northern lights have swirled across boreal skies for as long as the forest has lived, beyond the reach and control of humanity. In contrast, wolves and their kin are easily within our reach and control, and distressingly vulnerable. In the twenty-first century, humankind, as always, will decide which animals live and which will not, which forests will stand and which will be felled. My only hope is that our wisdom matches our capacity to control the destiny of our natural world.

Afterword

Threats to the Boreal Forest

By **Dr. Fiona K. A. Schmiegelow**, Conservation Biologist, University of Alberta, Edmonton

As a doctoral student, I aspired to study the effects of forest loss and fragmentation on wildlife communities by conducting a landscape experiment. After considering potential sites, including the spectacular red and white pine forests of the Temagami region of northern Ontario and the poster-perfect coastal forests of Clayoquot Sound on Vancouver Island, I settled on the boreal forest of northern Alberta. Unenlightened on the subtle splendor of the taiga, I knew that conservation issues were pressing and that I would do sound science. Little did I know this was also the start of a great adventure, which continues to this day.

My first summer in the boreal was spent in the utopia of a natural system where the human footprint was relatively light. Mornings were heralded by a cacophony of bird song, and daily explorations revealed new wonders. One memorable evening yielded a surprise encounter with a lone wolf on a post-dinner stroll, the sight of a sandhill crane foraging on the shore opposite our campsite, and a pair of frantic red-necked grebes over whose nest hovered a bald eagle. I had watched these grebes as they courted, as they carefully constructed a floating nest, as the female laid her eggs, and as the pair faithfully incubated their investment. I watched, with equal fascination, as the eagle methodically consumed the contents of each egg. That was nearly ten years ago.

The intervening years have brought many changes to this area. In the first winter of my study, a new road extended access west, servicing gas pipelines and numerous wells. The asso-

ciated compressor station was expanded to include several outbuildings and a large flare stack, which lights up the surrounding forest as it burns off sour gas. Two years later the area was harvested, and large expanses of continuous forest were fragmented by a patchwork of clearcuts. Extensive exploration for oil and gas reserves cuts swaths through the remaining forest each winter, and additional well sites pop up with regularity. Shadowy operations associated with gold prospecting come and go. And this past winter, a phone call alerted me to new drilling associated with diamond exploration.

I recount this litany of activity because it is typical of what is happening throughout the expanse of the North American taiga. The rate and extent of development in this region is unprecedented and alarming, and the cumulative effects of these activities on boreal systems are not known. On top of it all, global climate change presents a pervasive threat. But what does this mean for boreal flora and fauna?

Plants and animals need habitat in sufficient quantity and quality to support local populations and to ensure the regional persistence of species. Virtually all human activities in the boreal forest reduce habitat size, either directly or indirectly, and many also affect the quality of the remaining habitat. This problem is not unique to the taiga, and it would be naive to think we can eliminate human disturbance from this ecosystem any more than we can remove it from any other ecosystem in the world. What differs, however, is the scale at which development is proceeding, and the public's awareness of these issues. Everyone knows that tropical forests are at risk and that most temperate forests have been severely fragmented for some time, but the vast and remote nature of the boreal forest renders the threats almost intangible. Also, for many the taiga is simply a frozen landscape cloaked in an expanse of stunted trees—not the sort of stuff that generally stirs souls.

Until recently, industrial development in the North American boreal forest was not widespread. Forestry activities were focused on a few coniferous species, and operations tended to be fairly localized. Energy developments for gas and oil extraction or production of hydroelectric power have only occurred in the past few decades. Although some mines have existed for many years, they are relatively few in number, and their direct effects again quite localized. Historically, perhaps the greatest incursion into the boreal came from the south, where clearing of land for agriculture converted forest into farmland. Nevertheless, these activities occurred largely on the fringe, and ready access into the core of the North American taiga simply did not exist. As recently as 1997, Canada was recognized by the World Resources Institute as containing 25 percent of Earth's remaining frontier forests—those forests considered large enough and sufficiently intact to sustain their flora and fauna indefinitely. Virtually all of these areas fall within the boreal region and suffer from rapid encroachment.

Technological innovation is fueled by resource demands, and the boreal forest is a showcase for this global waltz. Recent advances in wood processing and increased mechanization have greatly enhanced the attractiveness of the taiga to the forest industry. Most tree species that grow in the region now have commercial value, and industrial forestry enables the rapid harvest of large areas. Oil and gas reserves not previously considered economically viable now also have value, and layer by layer, are allocated and exploited in a free-market frenzy. Increasingly sophisticated geophysical technologies continue to reveal mineral riches below the forest floor. Power demands and perceptions of impending water shortages stimulate proposals for an increasing number of dam and water diversion projects.

The most widespread human disturbance in the boreal is forest harvesting. Approximately 50 percent of the North American boreal forest is under some sort of forest tenure arrangement. Given that only about half of the region supports productive forest, this represents almost full allocation for timber production. Many of these allocations have occurred within the past fifteen years, and it is not unusual for agreements with forest companies to cover millions of hectares of public land. The predominant method of harvest is clear-cut logging, and logging activities target older forests, where wood volumes are highest. These forests then enter an economic rotation, the time between subsequent harvests dependent on the tree species of primary interest. Such rotations do not provide for replacement of older forests.

In many areas of the boreal forest, the rationale provided for harvesting practices is to emulate fire, the dominant natural disturbance in this region. While this is an attractive proposition, the analogy is problematic. Harvesting is much more selective than fire with respect to

Traditionally, snowshoes were the primary method of travel used by boreal Native peoples in winter.

the age and composition of the forest consumed, and less variable in terms of the amount and distribution of material left standing afterwards. Recent work also suggests that forest conditions following harvest or fire may be quite different for up to thirty years. Even if rates of harvest do not exceed estimated rates of natural disturbance, large fires still burn. Thus, the overall disturbance rate is inevitably higher.

What effects are current harvest practices likely to have in the boreal forest? Simply put, the forest will become younger and more uniform. Wildlife and plant species dependent on older forests for their existence are likely to suffer population declines and may become locally extinct in some areas. Post-fire specialists, such as the black-backed woodpecker, are also at risk, as salvage logging of burned areas effectively converts them to clear-cuts. Although we suffer from a paucity of boreal research in North America, lists of species at risk from Sweden and Finland, both countries with long histories of harvesting in boreal forests, are telling.

In much of the western North American boreal forest, oil and gas developments run a close second to forestry with respect to the amount of forest cut. However, unlike forests harvested and managed for timber, forest loss associated with energy sector activities is generally more permanent. In addition, the linear nature of these disturbances results in greater habitat fragmentation for some species. Woodland caribou, for example, avoid roads, pipelines, seismic lines, and well sites in parts of their range. Female wolverines generally den in road-free areas. And some forest birds, including the black-capped chickadee and yellow-rumped warbler, are reluctant to cross openings in the forest of 30 meters (98 ft.) or more.

Consider the latest mega-pipeline project, a 2,988-kilometer (1,857-mile) line that, when completed, will stretch from northeastern British Columbia to Chicago, Illinois. Its mission is to carry natural gas from western Canada to the Chicago area, for distribution throughout North America. Although plans state that the pipeline will generally not exceed 32 meters (105 ft.) in width, in areas where it twins with existing developments, associated right-of-ways may be up to 100 meters (328 ft.) wide. Now imagine the secondary pipelines that will feed the mainline, their associated well sites, the seismic lines created to locate the reserves, and the roads necessary to service these activities. A highly structured network is unfolding on a natural landscape characterized by its mosaic nature.

A recurrence in the boreal forest, beyond its green mantle of trees, is the presence of water. Whether in mighty rivers, majestic lakes, or the plethora of wetland and peatland complexes, water is everywhere. An estimated 40 percent of the waterfowl that migrate through the prairie potholes region nest in the western boreal forest. On a fine scale, road crossings in the taiga channel water in unnatural ways, affecting the hydrology of the surrounding area and altering local vegetation patterns and wetland complexes. Poorly constructed culverts may also disrupt the movement of fish. On a grander scale, major dam and water diversion projects affect the hydrology of entire regions, and effluent discharges from pulp mills can seriously degrade water quality. There are approximately three hundred dams throughout the Canadian boreal forest, and pulp mills dot every major watercourse. Many water diversion projects have been proposed for the boreal forest—one involves funneling the Yukon, Peace, and Liard Rivers into the Rocky Mountain Trench, to create an 800-kilometer (497-mi.), transborder reservoir. Another would see diversion of part of the Mackenzie watershed into Lake Winnipeg, for export to the United States. Yet another proposes running an underwater pipeline from Alaska to California. At present, the economics of these proposals are not favorable, and public sentiment does not support such developments. However, both could change in the future.

The southern boundary of the boreal continues to be eroded by agricultural expansion and liquidation of private forests for direct economic gain. Mills to the north provide a ready market, and current tax systems do not favor woodlot

The diminutive boreal owl nests in old woodpecker holes and may raise up to six chicks.

The aloofness of the northern forest disappears as you embrace its magic, and one cannot help but pledge allegiance to ensuring its future.

Fireweed flourishes four to five years after a wildfire.

conservation. The looming threat of global warming may push this boundary ever northward. Projections suggest that climatic conditions will change more quickly than the rate at which trees and other flora and fauna can adapt. As grasslands replace forest, the boreal forest will shrink. Northward expansion is limited by soil availability and large water bodies. Climate warming is also expected to lower water tables in boreal lakes and change the distribution of surface and ground water.

The boreal forest is not a passive player in climate change. Trees convert carbon dioxide, a "greenhouse" gas, into woody tissue and leaves, effectively locking it away until the trees burn, decompose, or are harvested. Peatlands are also carbon sinks. Extraction of fossil fuels releases carbon, as do other disturbances. Global warming may result in an increased number of forest fires. Some speculate that the large number of disturbances in the boreal forest in recent decades may have changed it from a carbon sink to a carbon source.

Amid the seemingly insurmountable challenges facing the boreal forest, I am comforted by recent memories. Despite the developments in my long-term study area, I am still awakened by bird song, and I can savor the ethereal quality of the forest as shafts of sunlight penetrate the dawn mist.

In tours of forestry operations, I have seen some new and innovative harvesting practices being implemented, and heard commitments to ecosystem management being made at corporate levels. If market pressures increase, this trend will spread. Similarly, other extraction industries, such as the oil and gas sector, will respond to environmental concerns from consumers if these affect economic gains. Governments can also be held to task for various national and international commitments to sustainable development, biodiversity conservation, and emissions reductions. Will these things happen quickly enough to stay the prevailing winds of change? It depends on how much we care.

A 1999 report from the Canadian Senate Subcommittee on the Boreal Forest identifies the issues as complex and recognizes that remedial actions require a substantial transition period. The report also cautions that the window of opportunity for preserving all the values offered by the boreal forest is quickly closing. I still believe we have a greater chance of balancing demands in the North American boreal forest than anywhere else on Earth. Recent surges in research have led to rapid increases in knowledge on both the limits and resiliency of the system. And there is a growing awareness of lessons to be gained from boreal experiences beyond North America.

My hope also lies with changing public sentiment and the coming generation of conservationists. I have the privilege of teaching some of these future leaders, and their enthusiasm and commitment is inspiring. I also see the marvel of the boreal forest reflected in the face of my child, as she traipses through the woods with sheer joy, stopping to gaze at a butterfly or a colorful fungus, and poking piles of scat to discover their contents. The aloofness of the northern forest disappears as you embrace its magic, and one cannot help but pledge allegiance to ensuring its future.

References and Further Reading

Alderman, T. "It's a Nuisance." *Imperial Oil Review*, Vol. 49, No. 3 (1965): 6–10.

Barbour, Michael G., and William Dwight Billings, eds. *North American Terrestrial Vegetation*. Cambridge, Eng.: Cambridge University Press, 1988.

Bastedo, Jamie. *Shield Country: Life and Times of the Oldest Piece of the Planet*. Calgary, AB: Arctic Institute of North America, 1994.

Bennett, Doug, and Tim Tiner. *Up North: A Guide to Ontario's Wilderness from Blackflies to the Northern Lights*. Markham, ON: Reed Books, 1993.

Bergerud, Arthur T., and Michael W. Gratson, eds. *Adaptive Strategies and Population Ecology of Northern Grouse*. Minneapolis, MN: University of Minnesota Press, 1988.

Berthold, Peter. *Bird Migration: A General Survey*. New York, NY: Oxford University Press, 1993.

Berthold, Peter. *Control of Bird Migration*. New York, NY: Chapman & Hall, 1996.

Bossenmaier, Eugene F. *Mushrooms of the Boreal Forest*. Saskatoon, SK: University of Saskatchewan Press, 1997.

Bryant, Dirk, Daniel Nielsen, and Laura Tangley. *The Last Frontier Forests: Ecosystems and Economies on the Edge*. World Resources Institute, 1997. http://www.wri.org/wri

Buller, Reginald. *Researches on Fungi*, Vol. 2. N.p., 1958.

Calef, George. *Caribou and the Barren-lands*. Willowdale, ON: Firefly Books Ltd., 1981.

Carbyn, L. N., S. H. Fritts, and D. R. Seip, eds. *Ecology and Conservation of Wolves in a Changing World*. Edmonton, AB: Canadian Circumpolar Institute, 1995.

Carbyn, L. N., S. M. Oosenburg, and D. W. Anions. *Wolves, Bison and the Dynamics Related to the Peace-Athabasca Delta in Canada's Wood Buffalo National Park*. Edmonton, AB: Canadian Circumpolar Institute, 1993.

Carpenteri, Stephen D. *Osprey: The Fish Hawk*. Minocqua, WI: NorthWord Press, 1997.

Chapman, F. M. *Warblers of North America*. New York, NY: D. Appleton, 1907.

Choate, Ernest A. *The Dictionary of American Bird Names*. Boston, MA: Gambit, 1973.

Churchfield, Sara. *The Natural History of Shrews*. Ithaca, NY: Comstock Publishing Associates, 1990.

Dobie, Frank. *The Voice of the Coyote*. Boston, MA: Little Brown, 1949.

Eastman, John. *The Book of Swamp and Bog*. Mechanicsburg, PA: Stackpole Books, 1995.

Formozov, A. N. *Snowcover as an Integral Factor of the Environment and its Importance in the Ecology of Mammals and Birds,* English edition. Edmonton, AB: Boreal Institute for Northern Studies, 1964.

Forsyth, Adrian. *Mammals of North America*. Willowdale, ON: Firefly Books Ltd., 1999.

Franzmann, Albert W., and Charles C. Schwartz, eds. *Ecology and Management of the North American Moose*. Washington, DC: Smithsonian Institution Press, 1998.

French, Hugh M., and Olav Slaymaker. *Canada's Cold Environments*. Montreal, QC, and Kingston, ON: McGill-Queen's University Press, 1993.

Gawthrop, Daniel. *Vanishing Halo: Saving the Boreal Forest*. Vancouver, BC: Greystone Books, 1999.

Gerrard, Jon M., and Gary R. Bortolotti. *The Bald Eagle: Haunts and Habits of a Wilderness Monarch*. Washington, DC: Smithsonian Institute Press, 1988.

Gittleman, John L., ed. *Carnivore Behavior, Ecology and Evolution*. Ithaca, NY: Comstock Publishing Associates, 1989.

Halfpenny, James C., and Roy Douglas Ozanne. *Winter: An Ecological Handbook*. Boulder, CO: Johnson Books, 1989.

Henry, J. David. *Red Fox: The Catlike Canine*. Washington, DC: Smithsonian Institution Press, 1986.

Hudler, George W. *Magical Mushrooms, Mischievous Molds*. Princeton, NJ: Princeton University Press, 1998.

Johnsgard, Paul A. *Diving Birds of North America*. Lincoln, NE: University of Nebraska Press, 1987.

Johnsgard, Paul A. *The Grouse of the World*. Lincoln, NE: University of Nebraska Press, 1983.

Johnson, Derek, Linda Kershaw, Andy MacKinnon, and Jim Pojar. *Plants of the Western Boreal Forest & Aspen Parkland*. Edmonton, AB: Lone Pine Publishing, 1995.

Juniper, B. E., R. J. Robins, and D. M. Joel. *The Carnivorous Plants*. New York, NY: Academic Press, 1989.

Kelsall, J. P. *The Caribou*. Ottawa, ON: Queen's Printer, 1968.

Kilham, Lawrence. *Woodpeckers of Eastern North America*. New York, NY: Dover Publications Ltd., 1983.

Larsen, James A. *Ecology of the Northern Lowland Bogs and Conifer Forests*. New York, NY: Academic Press, 1982.

Larsen, James A. *The Boreal Ecosystem*. New York, NY: Academic Press Inc., 1980.

Lehane, M. J. *Biology of Blood-sucking Insects*. London, Eng.: Harper Collins Academic, 1991.

Lynch, Wayne. *Bears: Monarchs of the Northern Wilderness*. Vancouver, BC: Douglas & McIntyre, 1993.

Lynch, Wayne. *Mountain Bears*. Calgary, AB: Fifth House Limited, 1999.

Marchand, Peter J. *Life in the Cold: An Introduction to Winter Ecology*, 3rd ed. Hanover, NH: University Press of New England, 1996.

May, Phillip R. A., Paul Newman, Ada Hirschman, and Joaquin M. Fuster. "Woodpeckers and Head Injury." *Lancet* (February 28, 1976).

McIntyre, Judith W. *The Common Loon: Spirit of Northern Lakes*. Toronto, ON: Fitzhenry & Whiteside, 1988.

McQueen, Cyrus B. *The Field Guide to the Peat Mosses of Boreal North America*. Hanover, NH: University Press of New England, 1990.

Merritt, Joseph F. *Winter Ecology of Small Mammals*. Pittsburgh, PA: Carnegie Museum of Natural History, 1984.

Moon, Barbara. *The Canadian Shield*. Toronto, ON: Natural Science of Canada Ltd., 1970.

Morse, Douglass H. *American Warblers: An Ecological and Behavioral Perspective*. Cambridge, MA: Harvard University Press, 1989.

Novak, Milan, James A. Baker, Martyn E. Obbard, and Bruce Malloch, eds. *Wild Furbearer Management and Conservation in North America*. Toronto, ON: Ontario Trappers Association, 1987.

O'Donnel, Colin, and Jon Fjeldså, comp. *Grebes—Status Survey and Conservation Action Plan*. Gland, Switzerland: IUCN, 1997.

Pielou, E. C. *After the Ice Age: The Return of Life to Glaciated North America*. Chicago, IL: University of Chicago Press, 1991.

Pielou, E. C. *The World of Northern Evergreens*. Ithaca, NY: Comstock Publishing Associates, 1988.

Poole, Alan F. *Ospreys: A Natural and Unnatural History*. Cambridge, Eng.: Cambridge University Press, 1989.

Pratt, Larry, and Ian Urquhart. *The Last Great Forest*. Edmonton, AB: NeWest Publishers Ltd, 1994.

Pruitt, William O. *Boreal Ecology*. London, Eng.: Edward Arnold Limited, 1978.

Roze, Uldis. *The North American Porcupine*. Washington, DC: Smithsonian Institution Press, 1989.

Russell, Anthony P., and Aaron M. Bauer. *The Amphibians and Reptiles of Alberta: A Field Guide and Primer of Boreal Herpetology*, 2nd ed. Calgary, AB: University of Calgary Press, 2000.

Seton, E. T. *The Arctic Prairies*. New York, NY: Scribner's, 1911.

Smith, Susan M. *Black-capped Chickadee (Wild Bird Guides)*. Mechanicsburg, PA: Stackpole Books, 1997.

Smith, Susan M. *The Black-capped Chickadee: Behavioral Ecology and Natural History*. Ithaca, NY: Comstock Publishing, 1991.

Smith, Wynet, and Peter Lee, eds. *Canada's Forests at a Crossroads: An Assessment in the Year 2000*. Global Forest Watch Canada, World Resources Institute, 2000. http://www.globalforestwatch.org

Stalmaster, Mark V. *The Bald Eagle*. New York, NY: Universe Books, 1987.

Stebbins, Robert C., and Nathan W. Cohen. *A Natural History of Amphibians*. Princeton, NJ: Princeton University Press, 1995.

Steller, Georg Wilhelm. In *Insects Through the Seasons*. Cambridge, MA: Harvard University Press, 1996.

Vander Wall, Stephen B. *Food Hoarding in Animals*. Chicago, IL: University of Chicago Press, 1990.

Waldbauer, Gilbert. *The Birder's Bug Book*. Cambridge, MA: Harvard University Press, 1998.

Weidensaul, Scott. *Living on the Wind: Across the Hemisphere with Migratory Birds*. New York, NY: North Point Press, 1999.

White, S. E. *The Forest*. N.p., 1903.

Wilson, Don E., and Sue Ruff. *The Smithsonian Book of North American Mammals*. Washington, DC: Smithsonian Institution Press, 1999.

Index